Modeling Extinction

Modeling Extinction

M. E. J. Newman
Santa Fe Institute
Santa Fe, NM

R. G. Palmer
Duke University
Durham, NC

Santa Fe Institute
Studies in the Sciences of Complexity

OXFORD
UNIVERSITY PRESS

2003

OXFORD

UNIVERSITY PRESS

Oxford   New York
Auckland   Bangkok   Buenos Aires   Cape Town   Chennai
Dar es Salaam   Delhi   Hong Kong   Istanbul   Karachi   Kolkata
Kuala Lumpur   Madrid   Melbourne   Mexico City   Mumbai   Nairobi
São Paulo   Shanghai   Taipei   Tokyo   Toronto

Copyright © 2003 by M. E. J. Newman and R. G. Palmer

Published by Oxford University Press, Inc.
198 Madison Avenue, New York, New York 10016

www.oup.com

Oxford is a registered trademark of Oxford University Press

Library of Congress Cataloging-in-Publication Data
Newman, M. E. J. (Mark E. J.)
Modeling extinction / M. E. J. Newman, R. G. Palmer.
p. cm. — (Santa Fe Institute studies in the sciences of complexity)
Includes bibliographical references (p.).
ISBN 0-19-515946-2 (pbk.) — ISBN 0-19-515945-4 (cloth)
1. Extinction (Biology)—Statistical methods. 2. Extinction (Biology)—
Mathematical models. I. Palmer, Richard G. II. Title. III. Proceedings volume
in the Santa Fe Institute studies in the sciences of complexity
QH78 .N48 2002
576.8'4—dc21   2002070092

1 3 5 7 9 8 6 4 2

Printed in the United States of America
on acid-free paper

About the Santa Fe Institute

The *Santa Fe Institute* (SFI) is a private, independent, multidisciplinary research and education center, founded in 1984. Since its founding, SFI has devoted itself to creating a new kind of scientific research community, pursuing emerging science. Operating as a small, visiting institution, SFI seeks to catalyze new collaborative, multidisciplinary projects that break down the barriers between the traditional disciplines, to spread its ideas and methodologies to other individuals, and to encourage the practical applications of its results.

All titles from the *Santa Fe Institute Studies in the Sciences of Complexity* series will carry this imprint which is based on a Mimbres pottery design (circa A.D. 950–1150), drawn by Betsy Jones. The design was selected because the radiating feathers are evocative of the out-reach of the Santa Fe Institute Program to many disciplines and institutions.

Santa Fe Institute
Studies in the Sciences of Complexity

Lecture Notes Volume

Author	*Title*
Eric Bonabeau, Marco Dorigo, and Guy Theraulaz	Swarm Intelligence: From Natural to Artificial Systems
M. E. J. Newman R. G. Palmer	Modeling Extinction

Proceedings Volumes

Editor	*Title*
James H. Brown and Geoffrey B. West	Scaling in Biology
Timothy A. Kohler and George J. Gumerman	Dynamics in Human and Primate Societies
Lee A. Segel and Irun Cohen	Design Principles for the Immune System and Other Distributed Autonomous Systems
H. Randy Gimblett	Integrating Geographic Information Systems and Agent-Based Modeling Techniques
James P. Crutchfield and Peter Schuster	Evolutionary Dynamics: Exploring the Interplay of Selection, Accident, Neutrality, and Function

Contents

Preface

Both scientists and the world in general have long had a fascination with extinction. "Why did the dinosaurs die out?" is one of the first true scientific questions that many children ask. But it is not merely the dinosaurs that became extinct. Almost every species that has ever lived upon the Earth has become extinct. Most survived for less than ten million years, merely the blink of an eye in the three-billion-year history of life on the planet. Extinction is the ultimate fate of every species, and the prevalence of extinction has played a crucial role in shaping the biosphere that we see today.

Traditional studies of extinction have, however, tended to adopt quite a narrow focus. Typical studies examine the causes of extinction of particular species, or groups of species, or of prominent mass extinction events, of which there have been a number during life's history on Earth. But in the last decade or so, scientists have started to examine another approach to the patterns of evolution and extinction in the fossil record. This approach might be called "statistical paleontology," since it looks at large-scale patterns in the record and attempts to understand and model their average statistical features, rather than their detailed structure. Examples of the types of patterns that these studies examine are the distribution of the sizes of mass extinction events over time, the distribution of species lifetimes, or the apparent increase in the number of species alive over the last half of a billion years. In attempting to model these patterns, researchers have drawn on ideas not only from paleontology, but from evolutionary biology, ecology, physics, and applied mathematics, including fitness landscapes, competitive exclusion, interaction matrices, and self-organized criticality. This book presents a self-contained review of work in this field, including a summary of the important features of the fossil record and detailed discussions of many of the models of extinction which have been proposed.

This book grew out of a meeting on extinction modeling held at the Santa Fe Institute in 1998. The book is not intended to reflect directly what was said at that meeting, but we have certainly benefited enormously from our discussions with the participants, as well as from subsequent discussions with many colleagues. We would, in particular, like to thank Per Bak, Stefan Boettcher, Gunther Eble, Doug Erwin, Wim Hordijk, Stuart Kauffman, Tim Keitt, Erik van Nimwegen, Andreas Pedersen, David Raup, Jack Sepkoski, Paolo Sibani, Kim Sneppen, and Ricard Solé. Thanks are due also to Chris Adami, Gunther Eble, Doug Erwin, and Jack Sepkoski for providing data used in a number of the figures. Finally, the authors would like to thank the Santa Fe Institute for their hospitality, funding, and patience during the writing of the manuscript. It is hard to imagine a more delightful work environment, or one more conducive to the successful completion of a project such as this.

Mark Newman
Santa Fe Institute

Richard Palmer
Duke University

November, 2001

CHAPTER 1
Extinction in the Fossil Record

Of the estimated one to four billion species that have existed on the Earth since life first appeared here (Simpson 1952), less than 50 million are still alive today (May 1990). All the others became extinct, typically within about ten million years (My) of their first appearance.[1] It is clearly a question of some interest what the causes are of this high turnover, and much research has been devoted to the topic (see, for example, Raup (1991a) and Glen (1994) and references therein). Most of this work has focussed on the causes of extinction of individual species, or on the causes of identifiable mass extinction events, such as the end-Cretaceous event. However, a recent body of work has examined instead the statistical features of the history of extinction, using mathematical models of extinction processes and comparing their predictions with global properties of the fossil record. In this book we will study these models, describing their mathematical basis, the extinction mechanisms that they incorporate, and their predictions.

Before we start looking at the models however, we need to learn something about the trends in fossil and other data which they attempt to model. This is the topic of this introductory chapter. Those well versed in the large-scale patterns seen in the Phanerozoic fossil record may wish to skip or merely browse this chapter, passing on to chapter 2, where the discussion of the models begins.

---

[1]In the paleontological literature, the abbreviation "My" is commonly used for millions of years, while the abbreviation "Ma" is used for millions of years *ago*. We use both throughout this book, as appropriate.

1.1 CAUSES OF EXTINCTION

There are two primary colleges of thought about the causes of extinction. The traditional view, still held by most palaeontologists as well as many in other disciplines, is that extinction is the result of external stresses imposed on the ecosystem by the environment (Benton 1991; Hoffmann and Parsons 1991; Parsons 1993). There are indeed excellent arguments in favor of this viewpoint, since we have good evidence for particular exogenous causes for a number of major extinction events in the Earth's history, such as marine regression (sea-level drop) for the late-Permian event (Jablonski 1985; Hallam 1989), and bolide impact for the end-Cretaceous (Alvarez et al. 1980; Alvarez 1983, 1987). These explanations are by no means universally accepted (Glen 1994), but almost all of the alternatives are also exogenous in nature, ranging from the mundane (climate change [Stanley 1984, 1988], ocean anoxia [Wilde and Berry 1984]) to the exotic (volcanism [Duncan and Pyle 1988; Courtillot et al. 1988], tidal waves [Bourgeois et al. 1988], magnetic field reversal [Raup 1985; Loper et al. 1988], supernovae [Ellis and Schramm 1995]). There seems to be little disagreement that, whatever the causes of these mass extinction events, they are the result of some change in the environment. However, the mass extinction events account for only about 35 percent of the total extinction evident in the fossil record at the family level, and for the remaining 65 percent we have no firm evidence favoring one cause over another. Many believe, nonetheless, that all extinction can be accounted for by environmental stress on the ecosystem. The extreme point of view has been put forward (though not entirely seriously) by Raup (1992), who used statistical analyses of fossil extinction and of the effects of asteroid impact to show that, within the accuracy of our present data, it is conceivable that *all* terrestrial extinction has been caused by meteors and comets. This, however, is more a demonstration of the uncertainty in our present knowledge of the frequency of impacts and their biotic effects than a realistic theory.

At the other end of the scale, an increasing number of biologists and ecologists are supporting the idea that extinction has biotic causes—that extinction is a natural part of the dynamics of ecosystems and would take place regardless of any stresses arising from the environment. There is evidence in favor of this viewpoint also, although it is to a large extent anecdotal. Maynard Smith (1989) has given a variety of different examples of modern-day extinctions caused by species interactions, such as the effects of overzealous predators or the introduction of new competitors into formerly stable systems. Unfortunately, the fossil record tells us only about the existence of extinct species, and not about the causes of their extinction. If species became extinct for a variety of different reasons, therefore, then they are all necessarily lumped together in our data and hard to distinguish from one another. Moreover, extinction events with biotic causes usually involve no more than a handful of species at the most, and are therefore too small to be picked out over the "background" level of extinction in the fossil data, making it difficult to say with any certainty whether they constitute an important part of this background extinction. (The distinction between mass and background extinction events is discussed in more detail in section 1.2.2.1.)

The recent modeling work which is the primary focus of this book attempts to address questions about the causes of extinction by looking at statistical trends in the extinction record, such as the relative frequencies of large and small extinction events. Using models which make predictions about these trends and comparing the results against fossil and other data, we can judge whether the assumptions which go into the models are plausible. Some of the models which we discuss are based on purely biotic extinction mechanisms, others on abiotic ones, and still others on some mixture of the two. While the results of this work are by no means conclusive yet—there are a number of models based on different extinction mechanisms which agree moderately well with the data—there has been some encouraging progress, and it seems a promising line of research.

1.2 THE DATA

In this section we review the palaeontological data on extinction. We also discuss a number of other types of data that may have bearing on the models that we will be discussing.

1.2.1 FOSSIL DATA

The discovery and cataloging of fossils is a painstaking business, and the identification of a single new species is frequently the sole subject of a published article in the literature. The models with which we are here concerned, however, predict statistical trends in species extinction, origination, diversification, and so on. In order to study such statistical trends, a number of authors have therefore compiled databases of the origination and extinction times of species described in the literature. The two most widely used such databases are those of Sepkoski (1992) and of Benton (1993). Sepkoski's data are labeled by both genus and family, although the genus-level data are, at the time of writing, unpublished. The database contains entries for approximately forty thousand marine genera, primarily invertebrates, from about five thousand families. Marine invertebrates account for the largest part of the known fossil record, and if one is to focus one's attention in any single area, this is the obvious area to choose. Benton's database, by contrast, covers both marine and terrestrial biotas, though it does so only at the family level, containing data on some seven thousand families. The choice of taxonomic level in a compilation such as this is inevitably a compromise. Certainly we would like data at the finest level possible, and a few studies have even been attempted at the species level (e.g., Patterson and Fowler 1996). However, the accuracy with which we can determine the origination and extinction dates of a particular taxon depend on the number of fossil representatives of that taxon. In a taxon for which we have very few specimens, the chances of one of those specimens lying close to the taxon's extinction date are slim, so that our estimate of this date will tend to be early. This bias is known as the Signor–Lipps effect (Signor and Lipps 1982). The reverse phenomenon, sometimes humorously referred to as the "Lipps–Signor" effect, is seen in the origination times of taxa, which in general err on the late side in poorly represented taxa. By grouping fossil species into higher taxa, we can work with denser data sets which give more accurate estimates of origination and extinction dates, at the expense of throwing out any information which is specific to the lower taxonomic levels (Raup and Boyajian 1988). (Higher taxa do, however, suffer from a greater tendency to paraphyly—see the discussion of pseudoextinction in section 1.2.2.5.)

1.2.1.1 Biases in the Fossil Data. The times of origination and extinction of species are usually recorded to the nearest geological stage. Stages are intervals of geological time determined by stratigraphic methods, or in some cases by examination of the fossil species present. While this is a convenient and widely accepted method of dating, it presents a number of problems. First, the dates of the standard geological stages are not known accurately. They are determined mostly by interpolation between a few widely spaced calibration points, and even the timings of the major boundaries are still contested. In the widely used time scale of Harland et al. (1990), for example, the Vendian–Cambrian boundary, which approximately marks the beginning of the explosion of multi-cellular life, is set at around 625 million years ago (Ma). However, more recent results indicate that its date may be nearer 545 Ma, a fairly significant correction (Bowring et al. 1993).

Another problem, which is particularly annoying where studies of extinction are concerned, is that the stages are not of even lengths. There are 77 stages in the Phanerozoic (the interval from the start of the Cambrian till the present, from which virtually all the data are drawn) with a mean length of 7.3 My, but they range in length from about 1 My to 20 My.[2] If one is interested in calculating extinction rates, i.e., the number of species becoming extinct per unit time, then clearly one should divide the number dying out in each stage by the length of the stage. However, if, as many suppose, extinction occurs not in a gradual fashion, but in intense bursts, this can give erroneous results. A single large burst of extinction which happens to fall in a short stage, would give an anomalously high extinction rate, regardless of whether the average extinction rate was actually any higher than in surrounding times. Benton (1995), for example, has calculated familial extinction rates in this way and finds that the apparent largest mass extinction event in the Earth's history was the late Triassic event, which is measured to be 20 times the size of the end-Cretaceous one. This result is entirely

---

[2] As we mentioned above, most species become extinct within 10 My of their first appearance, so a substantial fraction of species turn over within each stage. Most of this turnover is due to "pseudoextinction" however—the replacement of a species by its own descendent species—and so it is not the case, as simple arithmetic might suggest, that there is a mass extinction event happening in every stage. Pseudoextinction is discussed in section 1.2.2.5.

an artifact of the short duration (1 to 2 My) of the Rhaetian stage at the end of the Triassic. In actual fact the late Triassic event killed only about half as many families as the end-Cretaceous. In order to minimize effects such as these, it has become common in studies of extinction to examine not only extinction rates (taxa becoming extinction per unit time) but also total extinction (taxa becoming extinct in each stage). While the total extinction does not suffer from large fluctuations in short stages as described above, it obviously gives a higher extinction figure in longer stages in a way which rate measures do not. However, some features of the extinction record are found to be independent of the measure used, and in this case it is probably safe to assume that they are real effects rather than artifacts of the variation in stage lengths.

The use of the stages as a time scale has other problems associated with it as well. For example, it appears to be quite common to assign a different name to specimens of the same species found before and after a major stage boundary (Raup and Boyajian 1988), with the result that stage boundaries "generate" extinctions—even species which did not become extinct during a mass extinction event may appear to do so, by virtue of being assigned a new name after the event.

There are many other shortcomings in the fossil record. Good discussions have been given by Raup (1979a), Raup and Boyajian (1988), and Sepkoski (1996). Here we just mention briefly a few of the most glaring problems. The "pull of the recent" is a name which refers to the fact that species diversity appears to increase toward recent times because recent fossils tend to be better preserved and easier to dig up. Whether this in fact accounts for all of the observed increase in diversity is an open question, one which we discuss further in section 1.2.2.3. A related phenomenon affecting recent species (or higher taxa) is that some of them are still alive today. Since our sampling of living species is much more complete than our sampling of fossil ones, this biases the recent record heavily in favor of living species. This bias can be corrected by removing living species from our fossil data.

The "monograph" effect is a source of significant bias in studies of taxon origination. The name refers to the apparent burst of speciation seen as the result of the work of one particularly zealous researcher or group of researchers investigating a particular period; the record will show a peak of speciation over a short period of geological time, but this is only because that period has been so extensively researched. A closely related phenomenon is the so-called "Lagerstätten" effect, which refers to the burst of speciation seen when the fruits of a particularly fossil-rich site are added to the database. These and other fluctuations in the number of taxa—the standing diversity—over geologic time can be partly corrected by measuring extinction as a fraction of diversity. Such "per taxon" measures of extinction however may miss real effects such as the slow increase in overall diversity over time discussed in section 1.2.2.3. For this reason it is common in fact to calculate both per taxon and actual extinction when looking for trends in fossil data. Along with the two ways of treating time described above, this gives us four different extinction "metrics": total number of taxa becoming extinct per stage, percentage of taxa becoming extinct per stage, number per unit time, and percentage per unit time.

A source of bias in measures of the sizes of mass extinction events is poor preservation of fossils after a large event because of environmental disturbance. It is believed that many large extinction events are caused by environmental changes, and that these same changes may upset the depositional regime under which organisms are fossilized. In some cases this results in the poor representation of species which actually survived the extinction event perfectly well, thereby exaggerating the measured size of the event. There are a number of examples of so-called Lazarus taxa (Flessa and Jablonski 1983) which appear to become extinct for exactly this reason, only to reappear a few stages later. On the other hand, the Signor–Lipps effect discussed above tends to bias results in the opposite direction. Since it is unlikely that the last representative of a poorly represented taxon will be found very close to the actual date of a mass-extinction event, it sometimes appears that species are dying out for a number of stages before the event itself, even if this is not in fact the case. Thus extinction events tend to get "smeared" backward in time. In fact, the existence of Lazarus taxa can help us to estimate the magnitude of this problem, since the Signor–Lipps effect should apply to these taxa also, even though we know that they existed right up until the extinction event (and indeed beyond).

With all these biases present in the fossil data, one may well wonder whether it is possible to extract any information at all from the fossil record about the kinds of statistical trends with which our models are concerned. However, many studies have been performed which attempt to eliminate one or more of these biases, and some results are common to all studies. This has been taken as an indication that at least some of the trends visible in the fossil record transcend the rather large error bars on their measurement. In the next section we discuss some of these trends, particularly those which have been used as the basis for models of extinction, or cited as data in favor of such models.

1.2.2   TRENDS IN THE FOSSIL DATA

There are a number of general trends visible in the fossil data. Good discussions have been given by Raup (1986) and by Benton (1995). Here we discuss some of the most important points, as they relate to the models with which this book is concerned.

1.2.2.1   Extinction Rates.
In figure 1.1 we show a plot of the number of families of marine organisms becoming extinct in each geological stage since the start of the Phanerozoic. The data are taken from an updated version of the compilation by Sepkoski (1992). It is clear from this plot that, even allowing for the irregular sizes of the stages discussed above, there is more variation in the extinction rate than could be explained by simple Poissonian fluctuations. In particular, a number of

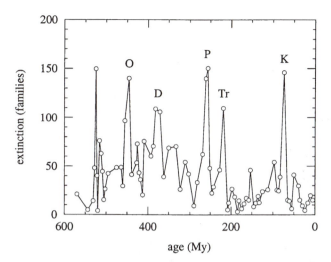

FIGURE 1.1　The number of families of known marine organisms becoming extinct per stratigraphic stage as a function of time during the Phanerozoic. The positions of the "big five" mass extinctions discussed in the text are marked with letters. The data are from the compilation by Sepkoski (1992).

mass extinction events can be seen in the data, in which a significant fraction of the known families were wiped out simultaneously. Palaeontology traditionally recognizes five large extinction events in terrestrial history, along with quite a number of smaller ones (Raup and Sepkoski 1982). The "big five" are led by the late Permian event (indicated by the letter P in the figure) which may have wiped out more than 90 percent of the species on the planet (Raup 1979b). The others are the events which ended the Ordovician (O), the Devonian (D), the Triassic (Tr), and the Cretaceous (K). A sixth extinction peak at about 525 Ma is also visible in the figure (the leftmost large peak), but it is still a matter of debate whether this peak represents a genuine historical event or just a sampling error; preservation is poor for this time period and the available data are much sparser than for the later extinction events.

　　As discussed in section 1.1, the cause of mass extinction events is a topic of much debate. However, it seems to be widely accepted that those causes, whatever they are, are abiotic, which lends strength to the view, held by many palaeontologists, that *all* extinction may have been caused by abiotic effects. The opposing view is that large extinction events may be abiotic in origin, but that smaller events, perhaps even at the level of single species, have biotic causes. Raup and Boyajian (1988) have investigated this question by comparing the extinction profiles of the nine major invertebrate groups throughout the Phanerozoic. While the similarities between these profiles are not as strong as between the extinction profiles of different subsets of the same group, they nonetheless

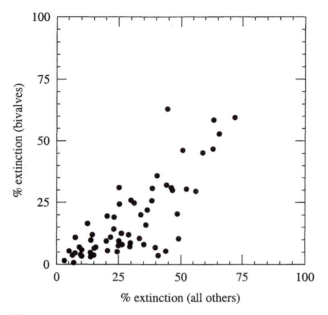

FIGURE 1.2   The percentage of genera of bivalves becoming extinct in each stage plotted against the percentage extinction of all other genera. The positive correlation ($r = 0.78$) indicates of a common cause of extinction. After Raup and Boyajian (1988).

find strong correlations between groups in the timing of extinction events. This may be taken as evidence that there is comparatively little taxonomic selectivity in the processes giving rise to mass extinction, which in turn favors abiotic rather than biotic causes. In figure 1.2, for example, reproduced from data given in their paper, we show the percentage extinction of bivalve families against percentage extinction of all other families, for each stage of the Phanerozoic. The positive correlation ($r^2 = 0.78$) of these data suggest a common cause for the extinction of bivalves and other species.

The shortcoming of these studies is that they can still only yield conclusions about correlations between extinction events large enough to be visible above the noise level in the data. It is perfectly reasonable to adopt the position that the large extinction events have exogenous causes, but that there is a certain level of "background" events which are endogenous in origin. In order to address this issue a number of researchers have constructed plots of the distribution of the sizes of extinction events; nonuniformity in such a distribution might offer support for distinct mass and background extinction mechanisms (Raup 1986; Kauffman 1993; Solé and Bascompte 1996). One such distribution is shown in figure 1.3, which is a histogram of the number of families dying out per stage. This is not strictly the same thing as the sizes of extinction events, since several

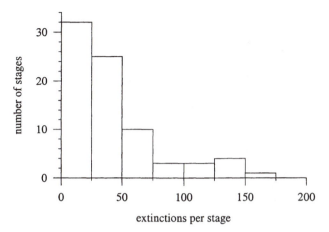

FIGURE 1.3   Histogram of the number of families of marine organisms becoming extinct per stratigraphic stage during the Phanerozoic. The data are drawn from Sepkoski (1992).

distinct events may contribute to the total in a given stage. However, since most extinction dates are only accurate to the nearest stage, it is the best we can do. If many independent extinction events were to occur in each stage, then one would expect, from Poisson statistics (see, for instance, Grimmett and Stirzaker 1992), that the histogram would be approximately normally distributed. In actual fact, as the figure makes clear, the distribution is highly skewed and very far from a normal distribution (Raup 1996). This may indicate that extinction at different times is correlated, with a characteristic correlation time of the same order of magnitude as or larger than the typical stage length so that the extinctions within a single stage are not independent events (Newman and Eble 1999b).

The histogram in figure 1.3 shows no visible discontinuities, within the sampling errors, and therefore gives no evidence for any distinction between mass and background extinction events. An equivalent result has been derived by Raup (1991b) who calculated a "kill curve" for marine extinctions in the Phanerozoic by comparing Monte Carlo calculations of genus survivorship with survivorship curves drawn from the fossil data. The kill curve is a cumulative frequency distribution of extinctions which measures the frequency with which one can expect extinction events of a certain magnitude. Clearly this curve contains the same information as the distribution of extinction sizes, and it can be shown that the conversion from one to the other involves only a simple integral transform (Newman 1996). On the basis of Raup's calculations, there is again no evidence for a separation between mass and background extinction events in the fossil record.

This result is not necessarily a stroke against extinction models which are based on biotic causes. First, it has been suggested (Jablonski 1986, 1991) that although there may be no quantitative distinction between mass and background events, there could be a qualitative one; it appears that the traits which confer survival advantages during periods of background extinction may be different from those which allow species to survive a mass extinction, so that the selection of species becoming extinction under the two regimes is different.

Second, there are a number of models which predict a smooth distribution of the sizes of extinction events all the way from the single species level up to the size of the entire ecosystem simply as a result of biotic interactions. In fact, the distribution of extinction sizes is one of the fundamental predictions of most of the models discussed in this book. Although the details vary, one of the most striking common features of these models is their prediction that the extinction distribution should follow a power law, at least for large extinction events. In other words, the probability $p(s)$ that a certain fraction s of the extant species/genera/families will become extinct in a certain time interval (or stage) should go like

$$p(s) \propto s^{-\tau}, \tag{1.1}$$

for large s, where τ is an exponent whose value is determined by the details of the model. This is a conjecture which we can test against the fossil record. In figure 1.4 we have replotted the data from figure 1.3 using logarithmic scales, on which a power-law form should appear as a straight line with slope $-\tau$. As pointed out by Solé and Bascompte (1996), and as we can see from the figure, the data are indeed compatible with the power-law form,[3] but the error bars are large enough that they are compatible with other forms as well, including the exponential shown in the figure.

In cases such as this, where the quality of the data makes it difficult to distinguish between competing forms for the distribution, a useful tool is the *rank/frequency plot*. A rank/frequency plot for extinction is constructed by taking the stratigraphic stages and numbering them in decreasing order of number of taxa becoming extinct. Thus the stage in which the largest number of taxa become extinct is given rank 1, the stage with the second largest number is given rank 2, and so forth. Then we plot the number of taxa becoming extinct as a function of rank. It is straightforward to show (Zipf 1949) that distributions which appear as power laws or exponentials in a histogram such as figure 1.4 will appear as power laws and exponentials on a rank/frequency plot also. However, the rank frequency plot has the significant advantage that the data points need not be grouped into bins as in the histogram. Binning the data reduces the number of independent points on the plot and throws away much of the information contained in our already sparse data set. Thus the rank/frequency plot often gives a better guide to the real form of a distribution.

In figure 1.5 we show a rank/frequency plot of extinctions of marine families in each stage of the Phanerozoic on logarithmic scales. As we can see, this plot

---

[3]In this case we have excluded the first point on the graph from our fit, which is justifiable since the power law is only expected for large values of s.

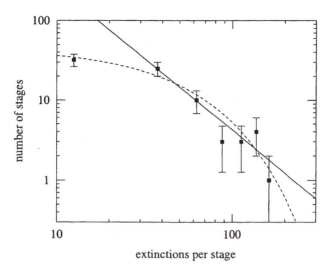

FIGURE 1.4   The data from figure 1.3 replotted on logarithmic scales, with Poissonian error bars. The solid line is the best power-law fit to the data. The dashed line is the best exponential fit.

does indeed provide a clearer picture of the behavior of the data, although ulti-mately the conclusions are rather similar. The points follow a power law quite well over the initial portion of the plot, up to extinctions on the order of 40 fam-ilies or so, but deviate markedly from power law beyond this point. The inset shows the same data with a linear time scale, and it appears that they may fall on quite a good straight line, although there are deviations in this case as well. Thus it appears again that the fossil data could indicate either a power-law or an exponential form (and possibly other forms as well).

More sophisticated analysis (Newman 1996) has not settled this question, although it does indicate that the Monte Carlo results of Raup (1991b) are in favor of the power-law form, rather than the exponential one, and also allows for a reasonably accurate measurement of the power law's exponent, giving $\tau = 2.0 \pm 0.2$. This value can be compared against the models' predictions.

1.2.2.2 Extinction Periodicity.   In an intriguing paper published in 1984, Raup and Sepkoski suggested that the mass extinction events seen in the most re-cent 250 My or so of the fossil record occur in a periodic fashion, with a period of about 26 My (Raup and Sepkoski 1984, 1986, 1988; Sepkoski 1989, 1990). Figure 1.6 shows the curve of extinction intensity for marine invertebrate genera from the middle Permian to the Recent from Sepkoski's data, with the postulated periodicity indicated by the vertical lines. A number of theories have been put forward, mostly based on astronomical causes, to explain how such a periodicity might arise (Davis et al. 1984; Rampino and Stothers 1984; Whitmire and Jack-

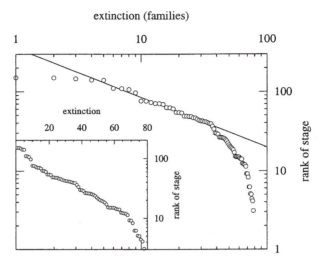

FIGURE 1.5 Main figure: a rank/frequency plot of the numbers of families of marine organisms becoming extinct in each stratigraphic stage of the Phanerozoic. The straight line is a power-law fit to the points. Inset: the same data replotted on semi-logarithmic axes.

son 1984; Hut et al. 1987). More recently however, it has been suggested that the periodicity has more mundane origins. Patterson and Smith (1987, 1989), for instance, have theorized that it may be an artifact of noise introduced into the data by poor taxonomic classification (Sepkoski and Kendrick [1993] argue otherwise), while Stanley (1990) has suggested that it may be a result of delayed recovery following large extinction events.

A quantitative test for periodicity of extinction is to calculate the power spectrum of extinction over the appropriate period and look for a peak at the frequency corresponding to 26 My. We have done this in figure 1.7 using data for marine families from the Sepkoski compilation. As the figure shows, there is a small peak in the spectrum around the relevant frequency (marked with an arrow), but it is not significant given the level of noise in the data. On the other hand, similar analyses by Raup and Sepkoski (1984) and by Fox (1987) using smaller databases do appear to produce a significant peak. The debate on this question is still in progress.

The power spectrum of fossil extinction is interesting for other reasons. Solé et al. (1997) have suggested on the basis of calculations using fossil data from the compilation by Benton (1993) that the spectrum has a $1/f$ form; i.e., it follows a power law with exponent -1. This result would be intriguing if true, since it would indicate that extinction at different times in the fossil record was correlated on arbitrarily long time scales. However, it now appears likely that the form found by Solé et al. is an artifact of the method of analysis, rather than

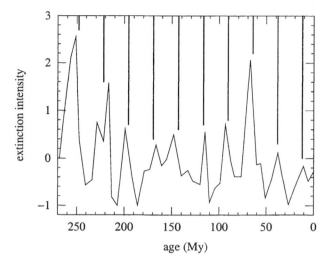

FIGURE 1.6 The number of genera of marine invertebrates becoming extinct per stratigraphic stage over the last 270 My. The vertical scale is in units of standard deviations from the mean extinction rate. The vertical bars indicate the positions of the periodic extinctions postulated by Raup and Sepkoski (1984). After Sepkoski (1990).

a real result (Kirchner and Weil 1998). Spectra calculated using other methods do not show the $1/f$ form and can be explained without assuming any long-time correlations: they are consistent with an exponential form at low frequencies crossing over to a $1/f^2$ behavior at high frequencies (Newman and Eble 1999b).

1.2.2.3 Origination and Diversity. The issue of origination rates of species in the fossil record is in a sense complementary to that of extinction rates, but has been investigated in somewhat less depth. Interesting studies have been carried out by, for example, Gilinsky and Bambach (1987), Jablonski and Bottjer (1990a, 1990b, 1990c), Jablonski (1993), Sepkoski (1998), and Eble (1998, 1999). One clear trend is that peaks in the origination rate appear in the immediate aftermath of large extinction events. In figure 1.8 we show the number of families of marine organisms appearing per stage. Comparison with figure 1.1 shows that there are peaks of origination corresponding to all of the prominent extinction peaks, although the correspondence between the two curves is by no means exact.

The usual explanation for these bursts of origination is that new species find it easier to get a toehold in the taxonomically underpopulated world which exists after a large extinction event. As the available niches in the ecosystem fill up, this is no longer the case, and origination slows. Many researchers have interpreted this to mean that there is a saturation level above which a given ecosystem can support no more new species, so that, apart from fluctuations in the immediate vicinity of the large extinction events, the number of species is

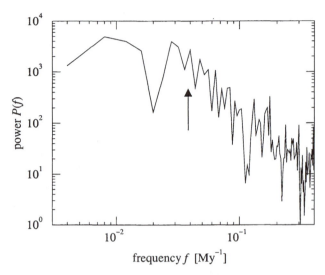

FIGURE 1.7   The power spectrum of familial extinction for marine animals over the last 250 My, calculated using data from the database compiled by Sepkoski (1992). The arrow marks the frequency corresponding to the conjectured 26 My periodicity of extinctions. Note that the scales on both axes are logarithmic.

approximately constant. This principle has been incorporated into most of the models considered in this book; the models assume a constant number of species and replace any which become extinct by an equal number of newly appearing ones. (The "reset" model considered in chapter 6 is an important exception.)

However, the hypothesis of constant species number is not universally accepted. In the short term, it appears to be approximately correct to say that a certain ecosystem can support a certain number of species. Modern-day ecological data on island biogeography support this view (see, for example, Rosenzweig [1995]). However, on longer time scales, the diversity of species on the planet appears to have been increasing, as organisms discover for the first time ways to exploit new habitats or resources. In figure 1.9 we show the total number of known fossil families as a function of geological time. The vertical axis is logarithmic, and the approximately straight-line form indicates that the increase in diversity is roughly exponential, although logistic and linear growth forms have been suggested as well (Sepkoski 1991; Newman and Sibani 1999). As discussed in section 1.2.1.1, one must be careful about the conclusions we draw from such figures, because of the apparent diversity increase caused by the "pull of the recent." However, current thinking mostly reflects the view that there is a genuine diversity increase toward recent times, associated with the expansion of life into new domains. As Benton (1995) has put it: "There is no evidence in the fossil record of a limit to the ultimate diversity of life on Earth."

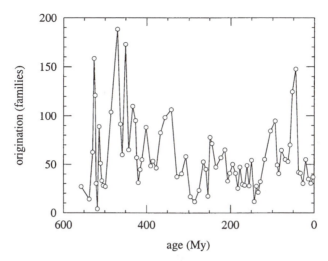

FIGURE 1.8 The number of families of known marine organisms appearing for the first time in each stratigraphic stage as a function of time throughout the Phanerozoic. The data come from the compilation by Sepkoski (1992).

1.2.2.4 Taxon Lifetimes. Another quantity which has been compared with the predictions of a variety of extinction models is the distribution of the lifetimes of taxa. In figure 1.10 we show a histogram of the lifetimes of marine genera in the Sepkoski database. The figure's axes are logarithmic and the solid and dotted lines represent respectively power-law and exponential fits to the data.

At first glance it appears from this figure that the lifetime distribution is better fitted by the exponential form. This exponential has a time constant of 40.1 My, which is of the same order of magnitude as the mean genus lifetime of 30.1 My. An exponential distribution of this type is precisely what one would expect to see if taxa are becoming extinct at random with a constant average rate (a Poisson process). A number of authors however have argued in favor of the power-law fit (Sneppen et al. 1995; Bak 1996). The power-law fit in the figure is a fit only to the data between 10 and 100 My. In this interval it actually matches the data quite well, but for longer or shorter lifetimes the agreement is poor. Why then should we take this suggestion seriously? The answer is that both very long and very short lifetimes are probably underrepresented in the database because of systematic biases. First, since the appearance and disappearance of genera are recorded only to the nearest stage, lifetimes of less than the length of the corresponding stage are registered as being zero and do not appear on the histogram. This means that lifetimes shorter than the average stage length of about 7 My are underrepresented. Second, as mentioned briefly in section 1.2.1.1, a taxon is sometimes given a different name before and after a major stage boundary, even though little or nothing about that taxon may have changed.

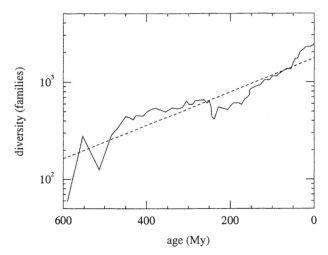

FIGURE 1.9   The total number of families of known fossil organisms as a function of time during the Phanerozoic. The vertical axis is logarithmic, and the dashed line is an exponential fit to the data. After Benton (1995).

This means that the number of species with lifetimes longer than the typical separation of these major boundaries is also underestimated in our histogram. This affects species with lifetimes greater than about 100 My. Thus there are plausible reasons for performing a fit only in the central region of figure 1.10 and, in this case, the power-law form is quite a sensible conjecture.

The exponent of the power law for the central region of the figure is measured to be $\alpha = 1.6 \pm 0.1$. This value is questionable however, since it depends on which data we choose to exclude at long and short times. In fact, a case can be made for any value between $\alpha = 1.2$ and 2.2. In this book we take a working figure of $\alpha = 1.7 \pm 0.3$ for comparison with theoretical models. Several of these models provide explanations for a power-law distribution of taxon lifetimes, with figures for α in reasonable agreement with this value.

We should point out that there is a very simple possible explanation for a power-law distribution of taxon lifetimes which does not rely on any detailed assumptions about the nature of evolution. If the addition and removal of species from a genus (or any subtaxa from a taxon) are stochastically constant and take place at roughly the same rate, then the number of species in the genus will perform an ordinary random walk. When this random walk reaches zero—the so-called first return time—the genus becomes extinct. Thus the distribution of the lifetimes of genera is also the distribution of first return times of a one-dimensional random walk. As is easily demonstrated (see Grimmett and Stirzaker [1992], for example), the distribution of first return times follows a power law with exponent 3/2, in reasonable agreement with the figure extracted from the fossil record above. An alternative theory is that speciation and extinction should

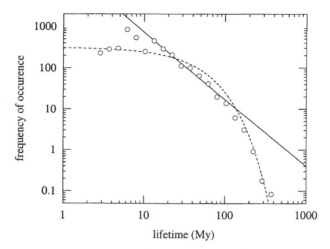

FIGURE 1.10   Frequency distribution of marine genus lifetimes in the fossil record. The solid line is the best power-law fit to the data between 10 and 100 My, while the dotted line is the best exponential fit to all the data. After Newman and Sibani (1999).

be multiplicative, i.e., proportional to the number of species in the genus. In this case the *logarithm* of the size of the genus performs a random walk, but the end result is the same: the distribution of lifetimes is a power law with exponent 3/2.

1.2.2.5 Pseudoextinction and Paraphyly.

One possible source of discrepancy between the models considered in this book and the fossil data is the way in which an extinction is defined. In the palaeontological literature a distinction is usually drawn between "true extinction" and "pseudoextinction." The term pseudoextinction refers to the evolution of a species into a new form, with the resultant disappearance of the ancestral form. The classic example is that of the dinosaurs. If, as is supposed by some, modern birds are the descendants of the dinosaurs (Gauthier 1986; Chiappe 1995), then the dinosaurs did not truly become extinct, but only pseudoextinct. Pseudoextinction is of course a normal part of the evolution process; Darwin's explanation of the origin of species is precisely the replacement of strains by their own fitter mutant offspring. And certainly this is a form of extinction, in that the ancestral strain will no longer appear in the fossil record. However, palaeontology makes a distinction between this process and true extinction—the disappearance of an entire branch of the phylogenetic tree without issue—presumably because the causes of the two are expected to be different. Pseudoextinction is undoubtedly a biotic process (although the evolution of a species and subsequent extinction of the parent strain may well be brought on by exogenous pressures—see Roy [1996], for example). On the other hand, many believe that we must look to environmental effects to find the causes of true extinction (Benton 1991).

Some of the models discussed in this book are models of true extinction, with species becoming extinct and being replaced by speciation from other, unrelated species. Others, however, deal primarily with pseudoextinction, predicting varying rates of evolution over the course of time, with mass extinction arising as the result of periods of intense evolutionary activity in which many species evolve to new forms, causing the pseudoextinction of their parent forms. It may not be strictly fair to compare models such as these to the fossil data on extinction presented above. To be sure, the data on extinction dates from which the statistics are drawn do not distinguish between true extinction and pseudoextinction; all that is recorded is the last date at which a specimen of a certain species is found. However, the grouping of the data into higher taxa, as discussed in section 1.2.1, does introduce such a distinction. When a species evolves to a new form, causing the pseudoextinction of the ancestral form, the new species is normally assigned to the same higher taxa—genus and family—as the ancestor. Thus a compilation of data at the genus or family level will not register the pseudoextinction of a species at this point. (This is reflected in the fact that the fraction of higher taxa that become extinct in a given period of time is almost always smaller than the fraction of species [Jablonski 1995].) The extinction of a genus or family can normally only occur when its very last constituent species becomes (truly) extinct, and therefore the data on the extinction times of higher taxa reflect primarily true extinctions.

However, the situation is not entirely straightforward. The assignment of species to genera and genera to families is, to a large extent, an arbitrary process, with the result that while the argument above may apply to a large portion of the data, there are many anomalies of taxonomy which give rise to exceptions. Strictly, the correct way to construct a taxonomic tree is to use cladistic principles. A clade is a group of species which all claim descendence from one ancestral species. In theory one can construct a tree in which each taxon is monophyletic, i.e., is composed only of members of one clade. Such a tree is not unique; there is still a degree of arbitrariness introduced by differences of opinion about when a species should be considered the founding member of a new taxon, and at what taxonomic level that taxon should be placed. However, to the extent that such species are a small fraction of the total, the arguments given above for the absence of pseudoextinction from the fossil statistics, at the genus level and above, are valid. In practice, however, cladistic principles are hard to apply to fossil species, whose taxonomic classification is based on morphology rather than on a direct knowledge of their lines of descent. In addition, a large part of our present classification scheme has been handed down to us by a tradition which predates the introduction of cladism. The distinction between dinosaurs and birds, for example, constitutes exactly such a traditional division. As a result, many—indeed most—taxonomic groups, particularly higher ones, tend to be paraphyletic: the members of the taxa are descended from more than one distinct ancestral species, whose own common ancestor belonged to another taxon. Not only does this failing upset our arguments concerning pseudoextinction above, but also, by virtue of the resulting unpredictable nature of the taxonomic hierarchy, introduces errors into our statistical measures of extinction which are hard to quantify (Sepkoski and Kendrick 1993). As Raup and Boyajian (1988) put it: "If all paraphyletic groups were eliminated from taxonomy, extinction patterns would certainly change."

1.2.3 OTHER FORMS OF DATA

There are a few other forms of data which are of interest in connection with the models that we will be discussing. Chief amongst these are taxonomic data on modern species, and simulation data from so-called artificial life experiments.

1.2.3.1 Taxonomic Data.

As long ago as 1922, it was noted that if one takes the taxonomic hierarchy of current organisms, counts the number of species n_s in each genus, and makes a histogram of the number of genera n_g for each value of n_s, then the resulting graph has a form which closely follows a power law (Willis 1922; Williams 1944):

$$n_g \sim n_s^{-\beta} . \tag{1.2}$$

In figure 1.11, for example, we reproduce the results of Willis for the number of species per genus of flowering plants. The measured exponent in this case is $\beta = 1.5 \pm 0.1$. Recently, Burlando (1990, 1993) has extended these results to higher taxa, showing that the number of genera per family, families per order, and so forth, also follow power laws, suggesting that the taxonomic tree has a fractal structure, a result of some interest to those working on "critical" models of extinction (see section 3.5).

In certain cases, for example, if one makes the assumption that speciation and extinction rates are stochastically constant, it can be shown that the average number of species in a genus bears a power-law relation to the lifetime of the genus, in which case Willis's data are merely another consequence of the genus lifetime distribution discussed in section 1.2.2.4. Even if this is true however, these data are nonetheless important, since they are derived from a source entirely different from the ones we have so far considered, namely from living species rather than fossil ones.

Note that we need to be careful about the way these distributions are calculated. A histogram of genus sizes constructed using fossil data drawn from a long period of geologic time is not the same thing as one constructed from a snapshot of genera at a single point in time. A snapshot tends to favor longer-lived genera which also tend to be larger, and this produces a histogram with a lower exponent than if the data are drawn from a long time period. Most of the models discussed in this book deal with long periods of geologic time and therefore mimic data of the latter kind better than those of the former. Willis's data, which are taken from living species, are inherently of the "snapshot" variety, and hence may have a lower value of β than that seen in fossil data and in models of extinction.

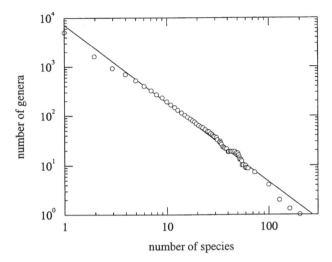

FIGURE 1.11   Histogram of the number of species per genus of flowering plants. The solid line is the best power-law fit to the data. After Willis (1922).

1.2.3.2 Artificial Life.   Artificial life (Langton 1995) is the name given to a class of evolutionary simulations which attempt to mimic the processes of natural selection, without imposing a particular selection regime from outside. (By contrast, most other computation techniques employing ideas drawn from evolutionary biology call upon the programmer to impose fitness functions or reproductive selection on the evolving population. Genetic algorithms [Mitchell 1996] are a good example of such techniques.) Probably the work of most relevance to the evolutionary biologist is that of Ray and collaborators (Ray 1994a, 1994b), who created a simulation environment known as Tierra, in which computer programs reproduce and compete for the computational resources of CPU time and memory. The basic idea behind Tierra is to create an initial "ancestor" program which makes copies of itself. The sole function of the program is to copy the instructions which comprise it into a new area of the computer's memory, so that, after some time has gone by, there will be a large number of copies of the same program running at once. However, the trick is that the system is set up so that the copies are made in an unreliable fashion. Sometimes a perfect copy is made, but sometimes a mistake occurs, so that the copy differs from the ancestor. Usually such mistakes result in a program which is not able to reproduce itself any further. However, occasionally they result in a program which reproduces more efficiently than its ancestor, and hence dominates over the ancestor after a number of generations. In systems such as this, many of the features of evolving biological systems have been observed, such as programs which cooperate in order to aid one another's efficient reproduction and parasitic programs which steal resources from others in order to reproduce more efficiently.

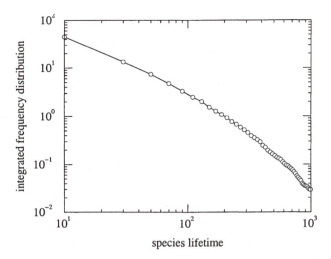

FIGURE 1.12   Plot of the integrated distribution of "species" lifetimes in runs of the Tierra artificial life simulation. The plot is approximately power-law in form except for a fall-off at large times due to the finite length of the runs. After Adami (1995).

In the context of the kinds of models that we will be studying here, the recent work of Adami (1995) using the Tierra system has attracted attention. In his work, Adami performed a number of lengthy runs of the Tierra simulation and observed the lifetimes of the species appearing throughout the course of the simulations. In figure 1.12 we show some of his results. The distribution of lifetimes appears again to follow a power law, except for a fall-off at long lifetimes, which may be explained by the finite length of the simulations.[4] This result appears to agree with the fossil evidence discussed in section 1.2.2.4, where the lifetimes of taxa were also found to follow a distribution approximately power-law in form. Possible explanations of this result have been discussed by Newman et al. (1997).

1.3   EARLY MODELS OF EXTINCTION

Most discussion of extinction has taken place at the species level, which is natural since extinction is intrinsically a species-level effect—by extinction we mean precisely the disappearance of a species, although the concept is frequently extended to cover higher taxa as well. Our discussion will also take place mostly at the species and higher taxonomic levels, but we should bear in mind that the processes underlying extinction occur at the level of the individual. McLaren (1988), for instance, has argued that it would be better to use the term "mass killing,"

---

[4]Although the integrated distribution in figure 1.12 does not appear to follow a straight line very closely, Adami (1995) shows that in fact it has precisely the form expected if the lifetimes are cut off exponentially.

rather than "mass extinction," since it is the death of individuals rather than species which is the fundamental process taking place.

Although many fossils of extinct species were unearthed during the eighteenth and early nineteenth centuries, it was not until the theory of evolution gained currency in the latter half of the nineteenth century that extinction became an accepted feature of the history of life on Earth.

One of the earliest serious attempts to model extinction was that of Lyell (1832) whose ideas, in some respects, still stand up even today. He proposed that when species first appear (he did not tackle the then vexed question of exactly how they appear) they possess varying fitnesses, and that those with the lowest fitness ultimately become extinct as a result of selection pressure from other species, and are then replaced by new species. While this model does not explain many of the most interesting features of the fossil record, it does already take a stand on a lot of the crucial issues in today's extinction debates: it is an equilibrium model with (perhaps) a roughly constant number of species and it has an explicit mechanism for extinction (species competition) which is still seriously considered as one of the causes of extinction. It also hints at of a way of quantifying the model by using a numerical fitness measure.

A few years after Lyell put forward his ideas about extinction, Darwin extended them by emphasizing the appearance of new species through speciation from existing ones. In his view, extinction arose as a result of competition between species and their descendants, and was therefore dominated by the process which we referred to as "pseudoextinction" in section 1.2.2.5. The Darwin–Lyell viewpoint is essentially a gradualist one. Species change gradually, and become extinct one by one as they are superseded by new fitter variants. As Darwin wrote in the *Origin of Species* (Darwin 1859): "Species and groups of species gradually disappear, one after another, first from one spot, then from another, and finally from the world." The obvious problem with this theory is the regular occurrence of mass extinctions in the fossil record. Although the existence of mass extinctions was well known in Darwin's time, Darwin and Lyell both argued strongly that they were probably a sampling artifact generated by the inaccuracy of dating techniques rather than a real effect. Today we know this not to be the case, and a purely gradualist picture no longer offers an adequate explanation of the facts. Any serious model of extinction must take mass extinction into account.

With the advent of reasonably comprehensive databases of fossil species, as well as computers to aid in their analysis, a number of simple models designed to help interpret and understand extinction data were put forward in the 1970s and 1980s. In 1973, van Valen proposed what he called the "Red Queen hypothesis": The probability per unit time of a particular species becoming extinct is independent of time. This "stochastically constant" extinction is equivalent to saying that the probability of a species surviving for a certain length of time t decays exponentially with t. This is easy to see, since if p is the constant probability per unit time of the species becoming extinct, then $1 - p$ is the probability that

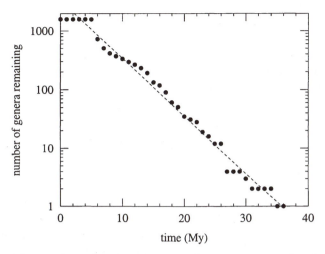

FIGURE 1.13   The number of genera of mammals surviving out of an initial group of 1585, over a period of 36 My. The dotted line is the best fit exponential, and has a time constant of 4.41 ± 0.08 My. After van Valen (1973).

it does not become extinct in any unit time interval, and

$$P(t) = (1 - p)^t = e^{-t/\tau} \tag{1.3}$$

is the probability that it survives t consecutive time intervals, where

$$\tau = -\frac{1}{\log(1 - p)} \simeq \frac{1}{p}, \tag{1.4}$$

where the second relation applies for small p. Van Valen used this argument to validate his hypothesis, by plotting "survivorship curves" for many different groups of species (van Valen 1973). A survivorship curve is a plot of the number of species surviving out of an initial group as a function of time starting from some arbitrary origin. In other words, one takes a group of species and counts how many of them are still present in the fossil record after time t. It appears that the time constant τ is different for the different groups of organisms examined by van Valen but roughly constant within groups, and in this case the survivorship curves should fall off exponentially. In figure 1.13 we reproduce van Valen's results for extinct genera of mammals. The approximately straight-line form of the survivorship curve on semilogarithmic scales indicates that the curve is indeed exponential, a result now known as "van Valen's law." Van Valen constructed similar plots for many other groups of genera and families and found similar stochastically constant extinction there as well.

Van Valen's result, that extinction is uniform in time has been used as the basis for a number of other simple extinction models, some of which are discussed in this book. However, for a number of reasons, it must certainly be incorrect. First, it is not mathematically possible for van Valen's law to be obeyed at more than one taxonomic level. As Raup (1991b) has demonstrated, if species become extinct at a stochastically constant rate p, the survivorship curve S for *genera* will not in general be exponential, because it depends not only on the extinction rate but also on the speciation rate. The general form for the genus survivorship curve is

$$S = 1 - \frac{p[e^{(q-p)t} - 1]}{qe^{(q-p)t} - p} \, , \tag{1.5}$$

where q is the average rate of speciation within the genus. A similar form applies for higher taxa as well.

Second, van Valen's law clearly cannot tell the whole story since, just like the theories of Lyell and Darwin, it is a gradualist model and takes no account of known mass extinction events in the fossil record. Raup (1991b, 1996) gives the appropriate generalization of van Valen's work to the case in which extinction is not stochastically constant. In this case, the *mean* survivorship curve follows van Valen's law (or eq. (1.5) for higher taxa), but individual curves show a dispersion around this mean whose width is a measure of the distribution of the sizes of extinction events. It was in this way that Raup extracted the kill curve discussed in section 1.2.2.1 for Phanerozoic marine invertebrates.

These models however, are all fundamentally just different ways of looking at empirical data. None of them offer actual explanations of the observed distributions of extinction events, or explain the various forms discussed in section 1.2. In the following chapters we discuss a variety of quantitative models which have been proposed in the last ten years to address these questions.

CHAPTER 2
Fitness Landscape Models

Kauffman (1993, 1995; Kauffman and Levin 1987; Kauffman and Johnsen 1991) has proposed and studied in depth a class of models referred to as NK models, which are models of random fitness landscapes on which one can implement a variety of types of evolutionary dynamics and study the development and interaction of species. (The letters N and K do not stand for anything; they are the names of parameters in the model.) Based on the results of extensive simulations of NK models, Kauffman and co-workers have suggested a number of possible connections between the dynamics of evolution and the extinction rate. To a large extent it is this work which has sparked recent interest in biotic mechanisms for mass extinction. In this chapter we review Kauffman's work in detail.

2.1 THE NK MODEL

An NK model is a model of a single rugged landscape, which is similar in construction to the spin-glass models of statistical physics (Fischer and Hertz 1991), particularly p-spin models (Derrida 1980) and random energy models (Derrida 1981). Used as a model of species fitness[5] the NK model maps the states

---

[5]NK models have been used as models of a number of other things as well—see, for instance, Kauffman and Weinberger (1989) and Kauffman and Perelson (1990).

genotype	w_1	w_2	w_3	W
000	0.487	0.076	0.964	0.509
001	0.851	0.372	0.398	0.540
010	0.487	0.097	0.162	0.249
011	0.851	0.566	0.062	0.493
100	0.235	0.076	0.964	0.425
101	0.311	0.372	0.398	0.360
110	0.235	0.097	0.162	0.165
111	0.311	0.566	0.062	0.313

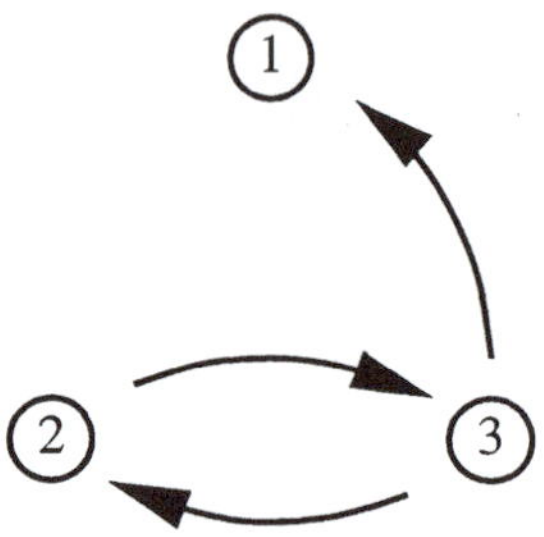

FIGURE 2.1   Calculation of the fitnesses for an NK model with three binary genes. In this case $K = 1$ with the epistatic interactions as indicated in the figure on the right.

of a model genome onto a scalar fitness W. This is a simplification of what happens in real life, where the genotype is first mapped onto phenotype and only then onto fitness. However, it is a useful simplification which makes simulation of the model for large systems tractable. As long as we bear in mind that this simplification has been made, the model can still teach us many useful things.

The NK model is a model of a genome with N genes. Each gene has A alleles. In most of Kauffman's studies of the model he used $A = 2$, a binary genetic code, but his results are not limited to this case. The model also includes epistatic interactions between genes—interactions whereby the state of one gene affects the contribution of another to the overall fitness of the species. In fact, it is these epistatic interactions which are responsible for the ruggedness of the fitness landscape. Without any interaction between genes it is possible (as we will see) to optimize individually the fitness contribution of each single gene, and hence to demonstrate that the landscape has the so-called Fujiyama form, with only a single global fitness peak.

In the simplest form of the NK model, each gene interacts epistatically with K others, which are chosen at random. The fitness contribution w_j of gene j is a function of the state of the gene itself and each of the K others with which it interacts. For each of the A^{K+1} possible states of these $K + 1$ genes, a value for w_j is chosen randomly from a uniform distribution between zero and one. The total fitness is then the average over all genes of their individual fitness contributions:

$$W = \frac{1}{N} \sum_{j=1}^{N} w_j . \tag{2.1}$$

This procedure is illustrated in figure 2.1 for a simple three-gene genome with $A = 2$ and $K = 1$.

Some points to notice about the NK model are:

1. The choices of the random numbers w_j are "quenched," which is to say that once they have been chosen they do not change again. The choices of the K

other genes with which a certain gene interacts are also quenched. Thus the fitness attributed to a particular genotype is the same every time we look at it.

2. There is no correlation between the contribution w_j of gene j to the total fitness for different alleles of the gene, or for different alleles of any of the genes with which it interacts. If any single one of these $K+1$ genes is changed to a different state, the new value of w_j is completely unrelated to its value before the change. This is an extreme case. In reality, epistatic interactions may have only a small effect on the fitness contribution of a gene. Again, however, this is a simplifying assumption which makes the model tractable.

3. In order to think of the NK model as generating a fitness "landscape" with peaks and valleys, we have to say which genotypes are close together and which far apart. In biological evolution, where the most common mutations are mutations of single genes, it makes sense to define the distance between two genotypes to be the number of genes by which they differ. This definition of distance, or "metric," is used in all the studies discussed here. A (local) peak is then a genotype that has higher fitness than all $N(A-1)$ of its nearest neighbors, those at distance 1 away.

4. The fact of taking an *average* over the fitness contributions of all the genes in eq. (2.1) is crucial to the behavior of the model. Taking the average has the effect that the typical height of fitness peaks diminishes with increasing N. In fact, one can imagine defining the model in a number of other ways. One could simply take the total fitness to be the sum of the contributions from all the genes—organisms with many genes therefore tending to be fitter than ones with fewer. In this case one would expect to see the reverse of the effect described above, with the average height of adaptive peaks increasing with increasing N. One might also note that since W is the sum of a number of independent random variables, its values should, by the central limit theorem, be approximately normally distributed with a standard deviation increasing as $\sqrt{N}$ with the number of genes. Therefore, it might make sense to normalize the sum with a factor of $N^{-1/2}$, so that the standard deviation remains constant as N is varied. Either of these choices would change some of the details of the model's behavior. For the moment however, we stick with the model as defined above.

What kind of landscapes does the NK model generate? Let us begin by considering two extreme cases. First, consider the case $K = 0$, in which all of the genes are entirely noninteracting. In this case, each gene contributes to the total fitness an amount w_j, which may take any of A values depending on the allele of the gene. The maximum fitness in this case is achieved by simply maximizing the contribution of each gene in turn, since their contributions are independent. Even if we assume an evolutionary dynamics of the most restrictive kind, in which we can only change the state of one gene at a time, we can reach the state of maximum fitness of the $K = 0$ model starting from any point on the landscape and only making changes which increase the fitness. Landscapes of this type are known as Fujiyama landscapes, after Japan's Mount Fuji: they are smooth and have a single global optimum.

Now consider the other extreme, in which K takes the largest possible value, $K = N - 1$. In this case each gene's contribution to the overall fitness W depends on itself and all $N - 1$ other genes in the genome. Thus if any single gene changes allele, the fitness contribution of every gene changes to a new random number, uncorrelated with its previous value. Thus the total fitness W is entirely uncorrelated between different states of the genome. This gives us the most rugged possible fitness landscape with many fitness peaks and valleys. The $K = N - 1$ model is identical to the random energy spin-glass model of Derrida (1981) and has been studied in some detail (Kauffman and Levin 1987; Macken and Perelson 1989). The fitness W in this case is the average of N independent uniform random variables between zero and one, which means that for large N it will be normally distributed about $W = 1/2$ with standard deviation $1/\sqrt{12N}$. This means that the typical height of the fitness peaks on the landscape decreases as $N^{-1/2}$ with increasing size of the genome. It also decreases with increasing K, since for larger K it is not possible to achieve the optimum fitness contribution of every gene, so that the average over all genes has a lower value than $K = 0$ case, even at the global optimum.

For values of K intermediate between the two extremes considered here, the landscapes generated by the NK model possess intermediate degrees of ruggedness. Small values of K produce highly correlated, smooth landscapes with a small number of high fitness peaks. High values of K produce more rugged landscapes with a larger number of lower peaks and less correlation between the fitnesses of similar genotypes.

2.2 EVOLUTION ON NK LANDSCAPES

In order to study the evolution of species using his NK landscapes, Kauffman made a number of simplifying assumptions. First, he assumed that evolution takes place entirely by the mutation of single genes, or small numbers of genes in an individual. That is, he neglected recombination. (This is a reasonable first approximation since, as we mentioned above, single gene mutations are the most common in biological evolution.) He also assumed that the mutation of different genes are *a priori* uncorrelated, that the rate at which genes mutate is the same for all genes, and that this rate is low compared to the time scale on which selection acts on the population. This last assumption means that the population can be approximated by a single genotype, and population dynamical effects can be ignored. (This may be valid for some populations, but is certainly not true in general.)

In addition to these assumptions it is also necessary to state how the selection process takes place, and Kauffman examined three specific possibilities, which he called the "random," "fitter," and "greedy" dynamics. If, as discussed above, evolution proceeds by the mutations of single genes, these three possibilities are as follows. In the random dynamics, single-gene mutations occur at random and, if the mutant genotype possesses a higher value of W than its ancestral strain, the mutant replaces the ancestor and the species "moves" on the landscape to the new genotype. A slight variation on this scheme is the fitter dynamics, in which a species examines all the genotypes which differ from the current genotype by the mutation of a single gene, its "neighbors," and then chooses a new genotype from these, either in proportion to fitness, or randomly amongst those which have higher fitness than the current genotype. (This last variation differs from the previous scheme only in a matter of time scale.) In the greedy dynamics, a species examines each of its neighbors in turn and chooses the one with the highest fitness W. Notice that whilst the random and fitter schemes are stochastic processes, the greedy one is deterministic; this gives rise to qualitative differences in the behavior of the model.

The generic behavior of the NK model of a single species is for the species' fitness to increase until it reaches a local fitness peak—a genotype with higher fitness than all of the neighboring genotypes on the landscape—at which point it stops evolving. For the $K = 0$ case considered above (the Fujiyama landscape), it will typically take on the order of N mutations to find the single fitness peak (or $N \log N$ for the random dynamics). For instance, in the $A = 2$ case, half of the alleles in a random initial genotype will on average be favorable and half unfavorable. Thus if evolution proceeds by the mutation of single genes, $1/2N$ mutations are necessary to reach the fitness maximum. In the other extreme, when $K = N - 1$, one can show that, starting from a random initial genotype, the number of directions which lead to higher fitness decreases by a constant factor at each step, so that the number of steps needed to reach one of the local maxima of fitness goes as $\log N$. For landscapes possessing intermediate values of K, the number of mutations needed to reach a local maximum lies somewhere between these limits. In other words, as N becomes large, the length of an adaptive walk to a fitness peak decreases sharply with increasing K. In fact, it appears to go approximately as $1/K$. This point will be important in our consideration of the many-species case. Recall also that the height of the typical fitness peak goes down with increasing K. Thus when K is high, a species does not have to evolve far to find a local fitness optimum, but in general that optimum is not very good.

2.3 COEVOLVING FITNESS LANDSCAPES

The real interest in NK landscapes arises when we consider the behavior of a number of coevolving species. Coevolution arises as a result of interactions between different species. The most common such interactions are predation, parasitism, competition for resources, and symbiosis. As a result of interactions such as these, the evolutionary adaptation of one species can prompt the adaptation of another (Vermeij 1987). Many examples are familiar to us, especially ones involving predatory or parasitic interactions. Plotnick and McKinney (1993) have given a number of examples of coevolution in fossil species, including predator-prey interactions between echinoids and gastropods (McNamara 1990) and mutualistic interactions between algae and foraminifera (Hallock 1985).

How is coevolution introduced into the NK model? Consider S species, each evolving on a different NK landscape. For the moment, let us take the simplest case in which each species has the same values of N and K, but the random fitnesses w_j defining the landscapes are different. Interaction between species is achieved by coupling their landscapes so that the genotype of one species affects the fitness of another. Following Kauffman and Johnsen (1991), we introduce two new quantities: S_i which is the number of neighboring species with which species i interacts,[6] and C which is the number of genes in each of those neighboring species which affect the fitness contribution of each gene in species i. On account of these two variables this variation of the model is sometimes referred to as the NKCS model.

Each gene in species i is "coupled" to C randomly chosen genes in each of the S_i neighboring species, so that, for example, if $C = 1$ and $S_i = 4$, each of i's genes is coupled to four other genes, one randomly chosen from each of four neighboring species. The coupling works in exactly the same way as the epistatic interactions of the last section—the fitness contribution w_j which a particular gene j makes to the total fitness of its host is now a function of the allele of that gene, of each of the K genes to which it is coupled *and* of the alleles of the CS_i genes in other species with which it interacts. As before, the values w_j are chosen randomly for each of the possible states of these genes.

The result is that when a species evolves so as to improve its own fitness, it may, in the process, change the allele of one of its genes which affects the fitness contribution of a gene in another species. As a result, the fitness of the other species will change. Clearly the further a species must evolve to find a fitness peak, the more alleles it changes, and the more likely it is to affect the fitness of its neighbors. Since the distance to a fitness peak depends on the value of K, so also does the chance of one species affecting another, and this is the root cause of the novel behavior seen in Kauffman's coevolution models.

The S_i neighboring species of species i can be chosen in a variety of different ways. The most common are either to chose them at random (but in a "quenched" fashion—once chosen, they remain fixed) or to place the species on a regular lattice, such as a square lattice in two dimensions, and then make the nearest neighbors of a species on the lattice its neighbors in the evolutionary sense.

---

[6]Although this quantity is denoted S_i, it is in fact a constant over all species in most of Kauffman's studies; the subscript i serves only to distinguish it from S, which is the total number of species. Of course, there is no reason why one cannot study a generalized model in which S_i (or indeed any of the other variables in the model, such as N or K) is varied from species to species, and Kauffman and Johnsen (1991) give some discussion and results for models of this type, although this is not their main focus.

In their original work on coevolving NK systems, Kauffman and Johnsen (1991) examined a number of different variations on the basic model outlined above. Here we consider the simplest case of relevance to extinction, the case of uniform K and S_i.

2.4 COEVOLUTIONARY AVALANCHES

Consider the case of binary genes ($A = 2$), with single-gene mutations. Starting from an initial random state, species take turns in strict rotation, and attempt by mutation to increase their own fitness irrespective of anyone else's. It is clear that if at any time all species in the system simultaneously find themselves at local fitness optima, then all evolution will stop, since there will be no further mutations of any species which can increase fitness. This state is known as a Nash equilibrium, a name taken from game theoretic models in which similar situations arise.[7] The fundamental question is whether such an equilibrium is ever reached. This, it turns out, depends on the value of K.

For large values of K, individual species landscapes are very rugged, and the distance that a species needs to go to reach a local fitness maximum is short. This means that the chance of it affecting its neighbors' fitness is rather small, and hence the chance of all species simultaneously finding a fitness maximum is quite good. On the other hand, if K is small, species must change many genes to reach a fitness maximum, and so the chances are high that they will affect the fitnesses of their neighbors. This in turn will force those neighbors to evolve, by moving the position of the maxima in their landscapes. They in turn may have to evolve a long way to find a new maximum, and this will affect still other species, resulting in an avalanche of coevolution which for small enough K never stops. Thus as K is decreased from large values to small, the typical size of the coevolutionary avalanche resulting from a random initial state increases until at some critical value K_c it becomes infinite.

What is this critical value? The product CS_i is the number of genes in other species on which the fitness contribution of a particular gene in species i depends. A rough estimate of the chance that at least one of these genes mutates during an avalanche is CS_iL, where L is the typical length of an adaptive walk of an isolated species (i.e., the number of genes which change in the process of evolving to a fitness peak). Assuming, as discussed in section 2.2, that L varies inversely with K, the critical value K_c at which the avalanche size diverges should vary as $K_c \sim CS_i$. This seems to be supported by numerical evidence: Kauffman and Johnsen found that $K_c \simeq CS_i$ in the particular case where every species is connected to every other ($S_i = S$).

---

[7]A related concept is that of the "evolutionarily stable strategy" (Maynard Smith and Price 1973), which is similar to a Nash equilibrium but also implies noninvadability at the individual level. The simulations of Kauffman and Johnsen considered here take place entirely at the species level, so "Nash equilibrium" is the appropriate nomenclature in this case.

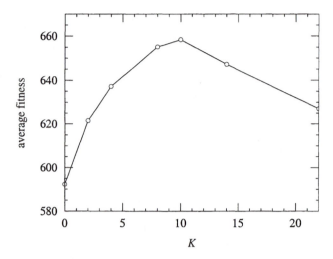

FIGURE 2.2   The average fitness of species in an NKCS model as a function of K. Twenty-five species were arranged in a 5×5 array so that each one interacted with $S_i = 4$ neighbors (except for those on the edges, for which $S_i = 3$, and those at the corners, for which $S_i = 2$). Each species had $N = 24$ and $C = 1$. The fitness plotted is that of the Nash equilibrium if reached, or the time average after transients if not. After Kauffman and Johnsen (1991).

The transition from the high-K "frozen" regime in which avalanches are finite to the low-K "chaotic" regime in which they run forever appears to be a continuous phase transition of the kind much studied in statistical physics (Binney et al. 1992). Bak et al. (1992) have analyzed this transition in some detail, showing that it does indeed possess genuine critical properties. Precisely at K_c, the distribution of the sizes s of the avalanches appears to be scale free and takes the form of a power law, eq. (1.1), which is typical of the "critical behavior" associated with such a phase transition. Kauffman and Johnson also pointed out that there are large fluctuations in the fitness of individual species near K_c, another characteristic of continuous phase transitions.

Figure 2.2 shows the average fitness of the coevolving species as a function of K for one particular case investigated by Kauffman and Johnsen. For ecosystems in the frozen $K > K_c$ regime the average fitness of the coevolving species increases from the initial random state until a Nash equilibrium is reached, at which point the fitness stops changing. As we pointed out earlier, the typical fitness of local optima increases with decreasing K, and this is reflected in the average fitness at Nash equilibria in the frozen phase: the average fitness of species at equilibrium increases as K approaches K_c from above.

In the chaotic $K < K_c$ regime a Nash equilibrium is never reached, but Kauffman and Johnsen measured the "mean sustained fitness," which is the average fitness of species over time, after an initial transient period in which the system settles down from its starting state. They found that this fitness measure *decreased* with decreasing K in the chaotic regime, presumably because species spend less and less time close to local fitness optima. Thus, there should be a maximum of the average fitness at the point $K = K_c$. This behavior is visible in figure 2.2, which shows a clear maximum around $K = 10$. The boundary between frozen and chaotic regimes was separately observed to occur at around $K_c = 10$ for this system.

On the basis of these observations, Kauffman and Johnsen then argued as follows. If the level of epistatic interactions in the genome is an evolvable property, just as the functions of individual genes are, and our species are able to "tune" the value of their own K parameter to achieve maximum fitness, then figure 2.2 suggests that they will tune it to the point $K = K_c$, which is precisely the critical point at which we expect to see a power-law distribution of coevolutionary avalanches. As we suggested in section 1.2.2.5, mass extinction could be caused by pseudoextinction processes in which a large number of species evolve to new forms nearly simultaneously. The coevolutionary avalanches of the NKCS model would presumably give rise to just such large-scale pseudoextinction. Another possibility, also noted by Kauffman and Johnson is that the large fluctuations in species fitness in the vicinity of K_c might be a cause of true extinction, low fitness species being more susceptible to extinction than high fitness ones.

These ideas are intriguing, since they suggest that by tuning itself to the point at which average fitness is maximized, the ecosystem also tunes itself precisely to the point at which species turnover is maximized, and indeed this species turnover is a large part of the reason why $K = K_c$ is a fit place to be in first place. Although extinction and pseudoextinction can certainly be caused by exogenous effects, even without these effects we should still see mass extinction.

Some efforts have been made to determine from the fossil evidence whether real evolution has a dynamics similar to the one proposed by Kauffman and co-workers. For example, Patterson and Fowler (1996) analyzed fossil data for planktic foraminifera using a variety of time-series techniques and concluded that the results were at least compatible with critical theories such as Kauffman's, and Solé et al. (1997) argued that the form of the extinction power spectrum may indicate an underlying critical macroevolutionary dynamics, although this latter suggestion has been questioned (Kirchner and Weil 1998; Newman and Eble 1999b).

2.5   COMPETITIVE REPLACEMENT

There is however a problem with the picture presented above. Numerically, it appears to be true that the average fitness of species in the model ecosystem is maximized when they all have K close to the critical value K_c. However, it would be a mistake to conclude that the system therefore must evolve to the critical point under the influence of selection pressure. Natural selection does not directly act to maximize the average fitness of species in the ecosystem, but rather it acts to increase individual fitnesses in a selfish fashion. Kauffman and Johnsen in fact performed simulations in which only two species coevolved, and they found that the fitness of both species was greater if the two had different values of K than if both had the value of K which maximized mean fitness. Thus, in a system in which many species could freely vary their own K under the influence of selection pressure, we would expect to find a range of K values, rather than all K taking the value K_c.

There are also some other problems with the original NKCS model. For instance, the values of K in the model were not actually allowed to vary during the simulations, but one would like to include this possibility. In addition, the mechanism by which extinction arises is rather vague; the model really only mimics evolution and the idea of extinction is tacked on somewhat as an afterthought.

To tackle all of these problems Kauffman and Neumann (unpublished) proposed a refinement of the NKCS model in which K can change and an explicit extinction mechanism is included, that of competitive replacement. (An account of this work can be found in Kauffman [1995].) In this variation of the model, a number S of species coevolve on NK fitness landscapes just as before. Now however, at each turn in the simulation, each species may change the state of one of its genes, it may change the value of its K by ± 1, it may be invaded by another species (see below), or it can do nothing. In their calculations, Kauffman and Neumann used the "greedy" dynamics described above and choose the change which most improves the fitness, but "fitter" and "random" variants are also possible. Allowing K to vary gives species the ability to evolve the ruggedness of their own landscapes in order to optimize their fitness.

Extinction takes place in the model when a species invades the niche occupied by another. If the invading species is better at exploiting the particular set of resources in the niche, it drives the niche's original occupant to extinction. In this model, a species' niche is determined by its neighboring species—the niche has no environmental component, such as climate, terrain, or food supply. Extinction by competitive replacement is actually not a very well-documented mode of extinction (Benton 1987). Maynard Smith (1989) has discussed the question at some length, but concludes that it is far more common for a species to adapt to the invasion of a new competitor than for it to become extinct. Nonetheless, there are examples of extinction by competitive replacement, and to the extent that it occurs, Kauffman and Neumann's work provides a model of the process. In the model, they add an extra "move" which can take place when a species' turn comes to evolve: it can be invaded by another species. A randomly chosen species can create a copy of itself (i.e., of its genome) which is then placed in the same niche as the first species and its fitness is calculated with respect to the genotypes of the neighbors in that niche. If this fitness exceeds the fitness of the original species in that niche, the invader supersedes the original occupant, which becomes extinct. In this way, fit species spread through the ecosystem making the average fitness over all species higher, but at the same time making the species more uniform, since over time the ecosystem will come to contain many copies of a small number of fit species, rather than a wide diversity of less-fit ones.

In numerical simulations this model shows a number of interesting features. First, regardless of their initial values, the Ks of the individual species appear to converge on an intermediate figure which puts all species close to the phase boundary discussed in the last section. This lends support to the ideas of Kauffman and Johnsen that fitness is optimized at this point (even though other arguments indicated that this might not be the best choice for selfishly evolving species—see above). Interestingly, the model also shows a power-law distribution of the sizes of extinction events taking place; if we count up the number of species becoming extinct at each time step in the simulation and make a histogram of these figures over the course of a long simulation, the result is of the form shown in figure 2.3. The power law has a measured exponent of $\tau \simeq 1$, which is not in good agreement with the figure of $\tau \simeq 2$ found in the fossil data (see section 1.2.2.1), but the mere existence of the power-law distribution is quite intriguing. Kauffman and Neumann explain its appearance as the result of avalanches of extinction which arise because the invasion of a niche by a new species (with the resulting extinction of the niche's previous occupant) disturbs the neighboring species, perhaps making them susceptible to invasion by further species. Another possible mechanism arises from the uniformity of genotypes

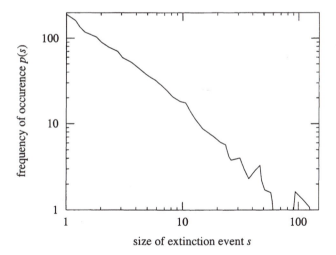

FIGURE 2.3   The distribution of the sizes of extinction events measured in a simulation of the model described in the text. The distribution is approximately power-law in form with an exponent measured to be $\tau = 1.18 \pm 0.03$. After Kauffman (1995).

which the invasion mechanism produces. As noted above, the invasion of many niches by one particularly fit species tends to produce an ecosystem with many similar species in it. If a new species arises which is able to compete successfully with these many similar species, then they may all become extinct over a short period of time, resulting in an extinction avalanche.

Why avalanches such as these should possess a power-law distribution is not clear. Kauffman and Neumann connect the phenomenon with the apparent adaptation of the ecosystem to the phase boundary between the ordered and chaotic regimes—the "edge of chaos" as Kauffman has called it. A more general explanation may come from the study of "self-organized critical" systems, which is the topic of the next chapter.

Kauffman and Neumann did not take the intermediate step of simulating a system in which species are permitted to vary their values of K, but in which there is no invasion mechanism. Such a study would be useful for clarifying the relative importance of the K-evolution and invasion mechanisms. Bak and Kauffman (unpublished, but discussed by Bak [1996]) have carried out some simulations along these lines, but apparently found no evidence for the evolution of the system to the critical point. Bak et al. (1992) have argued on theoretical grounds that such evolution should *not* occur in the maximally rugged case $K = N - 1$, but the argument does not extend to smaller values of K. In the general case the question has not been settled and deserves further study.

Self-Organized Critical Models

The models discussed in the last chapter are intriguing, but present a number of problems. In particular, most of the results about them come from computer simulations, and little is known analytically about their properties. Results such as the power-law distribution of extinction sizes and the system's evolution to the "edge of chaos" are only as accurate as the simulations in which they are observed. Moreover, it is not even clear what the mechanisms responsible for these results are, beyond the rather general arguments that we have already given. In order to address these shortcomings, Bak and Sneppen (1993; Sneppen et al. 1995; Sneppen 1995; Bak 1996) have taken Kauffman's ideas, with some modification, and used them to create a considerably simpler model of large-scale coevolution which also shows a power-law distribution of avalanche sizes and which is simple enough that its properties can, to some extent, be understood analytically. Although the model does not directly address the question of extinction, a number of authors have interpreted it, using arguments similar to those of section 1.2.2.5, as a possible model for extinction by biotic causes.

The Bak–Sneppen model is one of a class of models that show "self-organized criticality," which means that regardless of the state in which they start, they always tune themselves to a critical point of the type discussed in section 2.4, where power-law behavior is seen. We describe self-organized criticality in more detail in section 3.2. First, however, we describe the Bak–Sneppen model itself.

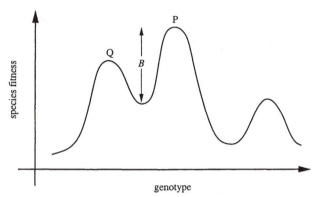

FIGURE 3.1 In order to reach a new adaptive peak Q from an initial genotype P, a species must pass through an intervening fitness "barrier," or region of low fitness. The height B of this barrier is a measure of how difficult it is for the species to reach the new peak.

3.1   THE BAK–SNEPPEN MODEL

In the model of Bak and Sneppen there are no explicit fitness landscapes, as there are in NK models. Instead the model attempts to mimic the effects of landscapes in terms of "fitness barriers." Consider figure 3.1, which is a toy representation of a fitness landscape in which there is only one dimension in the genotype (or phenotype) space. If the mutation rate is low compared with the time scale on which selection takes place (as Kauffman assumed), then a population will spend most of its time localized around a peak in the landscape (labeled P in the figure). In order to evolve to another, adjacent peak (Q), we must pass through an intervening "valley" of lower fitness. This valley presents a barrier to evolution because individuals with genotypes which fall in this region are selected against in favor of fitter individuals closer to P. In their model, Bak and Sneppen assumed that that the average time t taken to mutate across a fitness barrier of this type goes exponentially with the height B of the barrier:

$$t = t_0 e^{B/T} , \tag{3.1}$$

where t_0 and T are constants. The value of t_0 merely sets the time scale, and is not important. The parameter T, on the other hand, depends on the mutation rate in the population, and the assumption that mutation is low implies that T is small compared with the typical barrier heights B in the landscape. Equation (3.1) was proposed by analogy with the so-called Arrhenius law of statistical physics rather than by appealing to any biological principles, and in the case of evolution on a rugged fitness landscape it may well not be correct (see section 3.3). Nonetheless, as we will argue later, eq. (3.1) may still be a reasonable approximation to make.

Based on eq. (3.1), Bak and Sneppen then made a further assumption. If mutation rate (and hence T) is small, then the time scales t for crossing slightly different barriers may be widely separated. In this case a species' behavior is to a good approximation determined by the lowest barrier which it has to cross to get to another adaptive peak. If we have many species, then each species i will have some lowest barrier to mutation B_i, and the first to mutate to a new peak will be the one with the lowest value of B_i (the "lowest of the low," if you like). The Bak–Sneppen model assumes this to be the case and ignores all other barrier heights.

The dynamics of the model, which we now describe, have been justified in different ways, some of them more reasonable than others. Probably the most consistent is that given by Bak (private communication) which is as follows. In the model there are a fixed number N of species. Initially each species i is allotted a random number $0 \leq B_i < 1$ to represent the lowest barrier to mutation for that species. The model then consists of the repetition of two steps:

1. We assume that the species with the lowest barrier to mutation B_i is the first to mutate. In doing so it crosses a fitness barrier and finds its way to a new adaptive peak. From this new peak it will have some new lowest barrier for mutation. We represent this process in the model by finding the species with the lowest barrier and assigning it a new value $0 \leq B_i < 1$ at random.
2. We assume that each species is coupled to a number of neighbors. Bak and Sneppen called this number K. (The nomenclature is rather confusing; the variables N and K in the Bak–Sneppen model correspond to the variables S and S_i in the NK model.) When a species evolves, it will affect the fitness landscapes of its neighbors, presumably altering their barriers to mutation. We represent this by also assigning new random values $0 \leq B_i < 1$ for the K neighbors.

And that is all there is to the model. The neighbors of a species can be chosen in a variety of different ways, but the simplest is, as Kauffman and Johnsen (1991) also did, to put the species on a lattice and make the nearest neighbors on the lattice neighbors in the ecological sense. For example, on a one dimensional lattice—a line—each species has two neighbors and $K = 2$.

So what is special about this model? Well, let us consider what happens as we repeat the above steps many times. Initially the barrier variables are uniformly distributed over the interval between zero and one. If N is large, the lowest barrier will be close to zero. Suppose this lowest barrier B_i belongs to species i. We replace it with a new random value which is very likely to be higher than the old value. We also replace the barriers of the K neighbors of i with new random values. Suppose we are working on a one-dimensional lattice, so that these neighbors are species $i - 1$ and $i + 1$. The new barriers we choose for these two species are also very likely to be higher than B_i, although not necessarily higher than the old values of B_{i-1} and B_{i+1}. Thus, the steps (i) and (ii) will on average raise the value of the lowest barrier in the system, and will continue to do so as we repeat them again and again. This cannot continue forever however, since as the value of the lowest barrier in the system increases, it becomes less and less likely that it will be replaced with a new value which is higher. Figure 3.2 shows what happens in practice. The initial distribution of barriers gets eaten away from the bottom at first, resulting in a "gap" between zero and the height of the lowest barrier. After a time however, the distribution comes to equilibrium with a value of about 2/3 for the lowest barrier. (The actual figure is measured to be slightly over 2/3; the best available value at the time of writing is 0.66702 ± 0.00003 (Paczuski, Maslov, and Bak 1996).)

Now consider what happens when we make a move starting from a state which has a gap like this at the bottom end of the barrier height distribution. The species with the lowest barrier to mutation is right on the edge of the gap. We find this species and assign it and its K neighbors new random barrier values. There is a chance that at least one of these new values will lie in the gap, which necessarily makes it the lowest barrier in the system. Thus on the next step of the model, this species will be the one to evolve. We begin to see how avalanches appear in this model: there is a heightened chance that the next species to evolve will be one of the neighbors of the previous species. In biological terms the evolution of one species to a new adaptive peak changes the shape of the fitness landscapes of neighboring species, making them more likely to evolve too. The process continues, until, by chance, all new barrier values fall above the gap. In this case the next species to evolve will not, in general, be a neighbor of one of the other species taking part in the avalanche, and for this reason we declare it to be the first species in a new avalanche, the old avalanche being finished.

As the size of the gap increases, the typical length of an avalanche also increases, because the chances of a randomly chosen barrier falling in the gap in the distribution become larger. As we approach the equilibrium value $B_c = 0.667$ the mean avalanche size diverges, a typical sign of a self-organized critical system.

3.2 SELF-ORGANIZED CRITICALITY

So what exactly is self-organized criticality? The phenomenon was first studied by Bak, Tang, and Wiesenfeld (1987), who proposed what has now become the standard example of a self-organized critical (SOC) model, the self-organizing sandpile. Imagine a pile of sand which grows slowly as individual grains of sand are added to it one by one at random positions. As more sand is added, the height of the pile increases, and with it the steepness of the pile's sides. Started by single grains avalanches increase in size with steepness until at some point the pile is so steep that the avalanches become formally infinite in size, which is to say there is bulk transport of sand down the pile. This bulk transport, in turn, reduces the steepness of the pile so that subsequent avalanches are smaller.

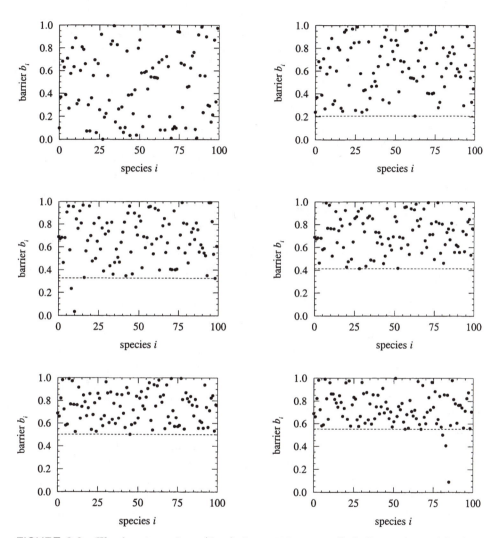

FIGURE 3.2   The barrier values (dots) for a 100-species Bak–Sneppen model after 50, 100, 200, 400, 800, and 1600 steps of a simulation. The dotted line in each frame represents the approximate position of the upper edge of the "gap" described in the text. In some frames a few species have barriers below this level, indicating that they were taking part in an avalanche at the moment when our snapshot of the system was taken.

The net result is that the pile "self-organizes" precisely to the point at which the infinite avalanche takes place, but never becomes any steeper than this.

A similar phenomenon takes place in the evolution model of Bak and Sneppen, and indeed the name "coevolutionary avalanche" is derived from the analogy between the two systems. The size of the gap in the Bak–Sneppen model plays the role of the steepness in the sandpile model. Initially, the gap increases as described above, and as it increases the avalanches become larger and larger on average, until we reach the critical point at which an infinite avalanche can occur. At this point the rates at which barriers are added and removed from the region below the gap exactly balance, and the gap stops growing, holding the system at the critical point thereafter.

It is interesting to compare the Bak–Sneppen model with the NKCS model discussed in section 2.3. Like the Bak–Sneppen model, the NKCS model also has a critical state in which power-law distributions of avalanches occur, but it does not self-organize to that state. It can be critical, but not self-organized critical. However, the essence of both models is that the evolution of one species distorts the shape of the fitness landscape of another (represented by the barrier variables in the Bak–Sneppen case), thus sometimes causing it to evolve too. So what is the difference between the two? The crucial point seems to be that in the Bak–Sneppen case the species which evolves is the one with the smallest barrier to mutation. This choice ensures that the system is always driven towards criticality.

At first sight, one apparent problem with the Bak–Sneppen model is that the delineation of an "avalanche" seems somewhat arbitrary. However, the avalanches are actually quite well separated in time because of the exponential dependence of mutation time scale on barrier height given by eq. (3.1). As defined above, an avalanche is over when no species remain with a barrier B_i in the gap at the bottom of the barrier height distribution, and the time until the next avalanche then depends on the first barrier B_i above the gap. If the "temperature" parameter T is small, then the exponential in eq. (3.1) makes this interavalanche time much longer than typical duration of a single avalanche. If we make a plot of the activity of the Bak–Sneppen model as a function of "real" time, (i.e., time measured in the increments specified by eq. (3.1)), the result looks like figure 3.3. In this figure the avalanches in the system are clearly visible and are well separated in time.

One consequence of the divergence of the average avalanche size as the Bak–Sneppen model reaches the critical point is that the distribution of the sizes of coevolutionary avalanches becomes scale-free—the size scale which normally describes it diverges and we are left with a distribution which has no scale parameter. The only (continuous) scale-free distribution is the power law, eq. (1.1), and, as figure 3.4 shows, the measured distribution is indeed a power law. Although the model makes no specific predictions about extinction, its authors argued, as we have done in section 1.2.2.5, that large avalanches presumably give rise to large-scale pseudoextinction, and may also cause true extinction via

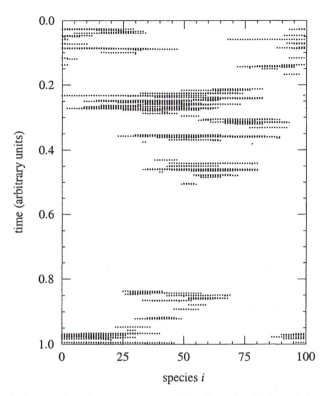

FIGURE 3.3   A time series of evolutionary activity in a simulation of the Bak–Sneppen model. Each dot represents the action of choosing a new barrier value for one species. Time in this figure runs down the page from top to bottom.

ecological interactions between species. They suggested that a power-law distribution of coevolutionary avalanches might give rise in turn to a power-law distribution of extinction events. The exponent τ of the power law generated by the Bak–Sneppen model lies strictly within the range $1 \leq \tau \leq 3/2$ (Bak and Sneppen 1993; Flyvbjerg et al. 1993), and if the same exponent describes the corresponding extinction distribution, this makes the model incompatible with the fossil data presented in section 1.2, which give $\tau \simeq 2$. However, since the connection between the coevolutionary avalanches and the extinction profile has not been made explicit, it is possible that the extinction distribution could be governed by a different, but related exponent which is closer to the measured value.

One of the elegant properties of SOC models, and critical systems in general, is that exponents such as τ above are *universal*. This means that the value of the exponent is independent of the details of the dynamics of the model, a point which has been emphasized by Bak (1996). Thus, although the Bak–Sneppen

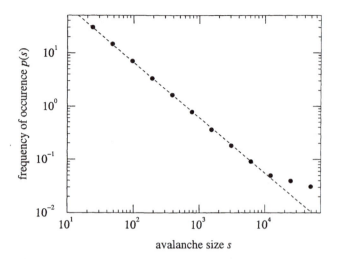

FIGURE 3.4 Histogram of the sizes of avalanches taking place in a simulation of an $N = 100$ Bak–Sneppen model on a one-dimensional lattice. The distribution is very close to power-law over a large part of the range, and the best-fit straight line (the dashed line above) gives a figure of $\tau = 1.04 \pm 0.01$ for the exponent.

model is undoubtedly an extremely simplified model of evolutionary processes, it may still be able to make quantitative predictions about real ecosystems, because the model and the real system share some universal properties.

3.3   TIME SCALES FOR CROSSING BARRIERS

Bak and Sneppen certainly make no claims that their model is intended to be a realistic model of coevolution, and therefore it may seem unfair to level detailed criticism at it. Nonetheless, a number of authors have pointed out shortcomings in the model, some of which have since been remedied by extending the model in various ways.

Probably the biggest criticism which can be leveled at the model is that the crucial eq. (3.1) is not a good approximation to the dynamics of species evolving on rugged landscapes. Weisbuch (1991) has studied this question in detail. He considers, as the models of Kauffman and of Bak and Sneppen also do, species evolving under the influence of selection and mutation on a rugged landscape in the limit where the rate of mutation is low compared with the time scale on which selection acts on populations. In this regime he demonstrates that the time scale t for mutation from one fitness peak across a barrier to another peak is given by

$$t = \frac{1}{qP_0} \prod_i \frac{F_0 - F_i}{q}, \qquad (3.2)$$

where q is the rate of mutation per gene, P_0 is the size of the population at the initial fitness peak, and F_i are the fitnesses of the mutant species at each genotype $i = 0, 1, 2, \ldots$ along the path in genotype space taken by the evolving species. The product over i is taken along this same path. Clearly this expression does not vary exponentially with the height of the fitness barrier separating the two fitness peaks. In fact, it goes approximately as a power law, with the exponent depending on the number of steps in the evolutionary path taken by the species. If this is the case, then the approximation implicit in eq. (3.1) breaks down and the dynamics of the Bak–Sneppen model is incorrect.

This certainly appears to be a worrying problem, but there may be a solution. Bak (1996) has suggested that the crucial point is that eq. (3.2) varies exponentially in the *number* of steps along the path from one species to another, i.e., the number of genes which must change to get us to a new genotype; in terms of the lengths of the evolutionary paths taken through genotype space, the time scales for mutation are exponentially distributed. The assumption that the "temperature" parameter T appearing in eq. (3.1) is small then corresponds to evolution which is dominated by short paths. In other words, mutations occur mostly between fitness peaks which are separated by changes in only a small number of genes. Whether this is in fact the case historically is unclear, though it is certainly well known that mutational mechanisms, such as recombination, which involve the simultaneous alteration of large numbers of genes are also an important factor in biological evolution.

3.4   THE EXACTLY SOLVABLE MULTITRAIT MODEL

The intriguing contrast between the simplicity of the rules defining the Bak–Sneppen model and the complexity of its behavior has led an extraordinary number of authors to publish analyses of its workings. (See Maslov et al. [1994]; de Boer et al. [1995]; Pang [1997]; and references therein for a subset of these publications.) In this chapter we will not delve into these mathematical developments in any depth, since our primary concern is extinction. However, there are several extensions of the model that *are* of interest to us. The first one is the "multitrait" model of Boettcher and Paczuski (1996). This model is a generalization of the Bak–Sneppen model in which a record is kept of several barrier heights for each species—barriers for mutation to different fitness peaks.

In the model of Boettcher and Paczuski, each of the N species has M independent barrier heights. These heights are initially chosen at random in the interval $0 \leq B < 1$. On each step of the model we search through all MN barriers to find the one which is lowest. We replace this one with a new value, and we also change the value of one randomly chosen barrier for each of the K neighboring species. Notice that the other $M - 1$ barrier variables for each species are left untouched. This seems a little strange; presumably if a species is mutating to a new fitness peak, all its barrier variables should change at once. However, the primary aim of Boettcher and Paczuski's model is not to mimic evolution more faithfully. The point is that their model is exactly solvable when $M = \infty$, which allows us to demonstrate certain properties of the model rigorously.

The exact solution is possible because, when $M = \infty$, the dynamics of the model separates into two distinct processes. As long as there are barrier variables whose values lie in the gap at the bottom of the barrier distribution, then the procedure of finding the lowest barrier will always choose a barrier in the gap. However, the second step—choosing at random one of the M barriers belonging to each of K neighbors—will *never* choose a barrier in the gap, since there are an infinite number of barriers for each species, and only ever a finite number in the gap. This separation of the processes taking place allowed Boettcher and Paczuski to write exact equations governing the dynamics of the system and to show that the model does indeed possess true critical behavior with a power-law distribution of avalanches.

The Bak–Sneppen model is the $M = 1$ limit of the multitrait generalization, and it would be very satisfying if it should turn out that the analytic results of Boettcher and Paczuski could be extended to this case, or indeed to any case of finite M. Unfortunately, no such extension has yet been found.

3.5 MODELS INCORPORATING SPECIATION

One of the other criticisms leveled at the Bak–Sneppen model is that it fails to incorporate speciation. When a biological population gives rise to a mutant individual which becomes the founder of a new species, the original population does not always die out. Fossil evidence indicates that it is common for both species to coexist for some time after such a speciation event. This process is absent from the Bak–Sneppen model, and in order to address this shortcoming Vandewalle and Ausloos (1995; Kramer et al. 1996) suggested an extension of the model in which species coexist on a phylogenetic tree structure, rather than on a lattice. The dynamics of their model is as follows.

Initially there is just a small number of species, perhaps only one, each possessing a barrier to mutation B_i whose value is chosen randomly in the range between zero and one. The species with the lowest barrier mutates first, but now both the original species and the mutant are assumed to survive, so that there is a branching of the tree leading to a pair of coexisting species (fig. 3.5). One might imagine that the original species should retain its barrier value, since this species is assumed not to have changed. However, if this were the case the model would never develop a "gap" as the Bak–Sneppen model does and so would never self-organize to a critical point. To avoid this, Vandewalle and Ausloos specified that both species, the parent and the offspring, should be assigned new, randomly chosen barrier values after the speciation event. We might justify this by saying for example that the environment of the parent species is altered by the presence of a closely related (and possibly competing)

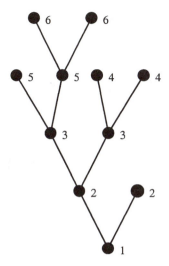

FIGURE 3.5 An example of a phylogenetic tree generated by the model of Vandewalle and Ausloos (1995). The numbers
indicate the order of growth of the tree.

offspring species, thereby changing the shape of the parent's fitness landscape.
Whatever the justification, the model gives rise to a branching phylogenetic tree
which contains a continuously increasing number of species, by contrast with the
other models we have examined so far, in which the number was fixed. As we
pointed out in section 1.2.2.3, the number of species in the fossil record does in
fact increase slowly over time, which may be regarded as partial justification for
the present approach.

In addition to the speciation process, there is also a second process taking
place, similar to that of the Bak–Sneppen model: after finding the species with
the lowest barrier to mutation, the barrier variables B_i of all species within a distance k of that species are also given new, randomly chosen values between zero
and one. Distances on the tree structure are measured as the number of straight-
line segments which one must traverse in order to get from one species to another
(see fig. 3.5 again). Notice that this means that the evolution of one species to a
new form is more likely to affect the fitness landscape of other species which are
closely related to it phylogenetically. There is some justification for this, since
closely related species tend to exploit similar resources and are therefore more
likely to be in competition with one another. On the other hand, predator-prey
and parasitic interactions are also very important in evolutionary dynamics, and
these interactions tend not to occur between closely related species.

Many of the basic predictions of the model of Vandewalle and Ausloos are
similar to those of the Bak–Sneppen model, indicating perhaps that Bak and
Sneppen were correct to ignore speciation events to begin with. It is found again
that avalanches of coevolution take place, and that the system organizes itself
to a critical state in which the distribution of the sizes of these avalanches follows a power law. The measured exponent of this power law is $\tau = 1.49 \pm 0.01$

(Vandewalle and Ausloos 1997), which is very close to the upper bound of $3/2$ calculated by Flyvbjerg et al. (1993) for the Bak–Sneppen model. However, there are also some interesting features which are new to this model. In particular, it is found that the phylogenetic trees produced by the model are self-similar. In section 1.2.3.1 we discussed the work of Burlando (1990), which appears to indicate that the taxonomic trees of living species are also self-similar. Burlando made estimates of the fractal (or Hausdorf) dimension D_H of taxonomic trees for 44 previously published catalogues of species taken from a wide range of taxa and geographic areas, and found values ranging from 1.1 to 2.1 with a mean of 1.6.[8] (The typical confidence interval for values of D_H was on the order of ± 0.2.) These figures are in reasonable agreement with the value of $D_H = 1.89 \pm 0.03$ measured by Vandewalle and Ausloos (1997) for their model, suggesting that a mechanism of the kind they describe could be responsible for the observed structure of taxonomic trees.

The model as described does not explicitly include extinction, and furthermore, since species are not replaced by their descendents as they are in the Bak–Sneppen model, there is also no pseudoextinction. However, Vandewalle and Ausloos also discuss a variation on the model in which extinction is explicitly introduced. In this variation, they find the species with the lowest barrier to mutation B_i and then they randomly choose either to have this species speciate with probability $1 - \exp(-B_i/r)$ or to have it become extinct with probability $\exp(-B_i/r)$, where r is a parameter which they choose. Thus the probability of extinction *decreases* with increasing height of the barrier. It is not at first clear how we are to understand this choice. Indeed, it seems likely from reading the papers of Vandewalle and Ausloos that there is some confusion between the barrier heights and the concept of fitness; the authors argue that the species with higher *fitness* should be less likely to become extinct, but then equate fitness with the barrier variables B_i. One way out of this problem may be to note that on rugged landscapes with bounded fitness there is a positive correlation between the heights of barriers and the fitness of species: the higher the fitness the more likely it is that the lowest barrier to mutation will also be high.

When $r = 0$, this extinction model is equal to the first model described, in which no extinction took place. When r is above some threshold value r_c, which is measured to be approximately 0.48 ± 0.01 for $k = 2$ (the only case the authors investigated in detail), the extinction rate exceeds the speciation rate and the tree ceases to grow after a short time. In the intervening range $0 < r < r_c$ evolution and extinction processes compete and the model shows interesting behavior. Again there is a power-law distribution of coevolutionary avalanches, and a fractal tree structure reminiscent of that seen in nature. In addition there is now a power-law distribution of extinction events, with the same exponent as the coevolutionary avalanches, i.e., close to $3/2$. As with the

---

[8]In fact, D_H is numerically equal to the exponent β for a plot such as that shown in figure 1.11 for the appropriate group of species.

Bak–Sneppen model this is in disagreement with the figure of 2.0 ± 0.2 extracted from the fossil data.

Another variation of the Bak–Sneppen model which incorporates speciation has been suggested by Head and Rodgers (1997). In this variation, they keep track of the two lowest barriers to mutation for each species, rather than just the single lowest. The mutation of a species proceeds in the same fashion as in the normal Bak–Sneppen model when one of these two barriers is significantly lower than the other. However, if the two barriers are close together in value, then the species may split and evolve in two different directions on the fitness landscape, resulting in speciation. How similar the barriers have to be in order for this to happen is controlled by a parameter δs, such that speciation takes place when

$$|B_1 - B_2| < \delta s, \tag{3.3}$$

where B_1 and B_2 are the two barrier heights. The model also incorporates an extinction mechanism, which, strangely, is based on the opposite assumption to the one made by Vandewalle and Ausloos. In the model of Head and Rodgers, extinction takes place when species have particularly *high* barriers to mutation. To be precise, a species becomes extinct if its neighbor mutates (which would normally change its fitness landscape and therefore its barrier variables) but both its barriers are above some predetermined threshold value. This extinction criterion seems a little surprising at first: if, as we suggested above, high barriers are positively correlated with high fitness, why should species with high barriers become extinct? The argument put forward by Head and Rodgers is that species with high barriers to mutation find it difficult to adapt to changes in their environment. To quote from their paper, "A species with only very large barriers against mutation has become so inflexible that it is no longer able to adapt and dies out." It seems odd however, that this extinction process should take place precisely in the species which are adjacent to others which are mutating. In the Bak–Sneppen model, these species have their barriers changed to new random values as a result of the change in their fitness landscapes brought about by the mutation of their neighbor. Thus, even if they did indeed have high barriers to mutation initially, their barriers would be changed when their neighbor mutated, curing this problem and so one would expect that these species would *not* become extinct.[9]

The model has other problems as well. One issue is that, because of the way the model is defined, it does not allow for the rescaling of time according to eq. (3.1). This means that evolution in the model proceeds at a uniform rate, rather than in avalanches as in the Bak–Sneppen model. As a direct result of this, the distribution of the sizes of extinction events in the model follows a Poisson distribution, rather than the approximate power law seen in the fossil data (fig. 1.3). The model does have the nice feature that the number of species

---

[9]A later paper on the model by Head and Rodgers (unpublished) has addressed this criticism to some extent.

in the model tends to a natural equilibrium; there is a balance between speciation and extinction events which causes the number of species to stabilize. This
contrasts with the Bak–Sneppen model (and indeed almost all the other models
we discuss) in which the number of species is artificially held constant, and also
with the model of Vandewalle and Ausloos, in which the number of species either
shrinks to zero, or grows indefinitely, depending on the value of the parameter r.
Head and Rodgers gave an approximate analytic explanation for their results using a "mean field" technique similar to that employed by Flyvbjerg et al. (1993)
for the Bak–Sneppen model. However, the question of whether the number of
species predicted by their model agrees with the known taxon carrying capacity
of real ecosystems has not been investigated.

3.6   MODELS INCORPORATING EXTERNAL STRESS

Another criticism of the approach taken in Bak and Sneppen's work (and indeed
in the work of Kauffman discussed in chapter 2) is that real ecosystems are not
closed dynamical systems, but are in reality affected by many external factors,
such as climate and geography. Indeed, as we discussed in section 1.2.2.1, a
number of the larger extinction events visible in the fossil record have been tied
quite convincingly to particular exogenous events, so that any model ignoring
these effects is necessarily incomplete. Newman and Roberts (1995; Roberts and
Newman 1996) have proposed a variation on the Bak–Sneppen model which
attempts to combine the ideas of extinction via environmental stress and large-
scale coevolution. The basic idea behind this model is that a large coevolutionary
avalanche will cause many species to move to new fitness peaks, some of which
may possess lower fitness than the peaks they previous occupied. Thus a large
avalanche produces a number of new species which have low fitness and therefore
may be more susceptible to extinction as a result of environmental stress. This in
fact is not a new idea. Kauffman, for example, has made this point clearly in his
book *The Origins of Order* (Kauffman 1993): "During coevolutionary avalanches,
species fall to lower fitness and hence are more likely to become extinct. Thus the
distribution of avalanche sizes may bear on the distribution of extinction events
in the fossil record."

   Newman and Roberts incorporated this idea into their model as follows.
A fixed number N of species each possess a barrier B_i to mutation, along with
another variable F_i which measures their fitness at the current adaptive peak. On
each step of the simulation the species with the lowest barrier B_i for mutation,
and its K neighbors, are selected, just as in the Bak–Sneppen model. The B_i and
F_i variables of these $K + 1$ species are all given new independent random values
between zero and one, representing the evolution of one species and the changed
landscapes of its neighbors. Then, a positive random number η is chosen which
represents the level of environmental stress at the current time, and all species

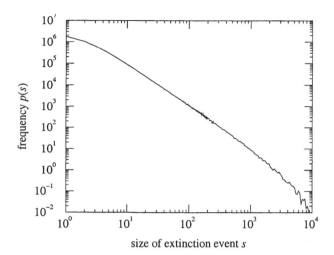

FIGURE 3.6 The distribution of sizes of extinction events in a simulation of the model of Newman and Roberts (1995) with $N = 10000$ and $K = 3$. The measured exponent of the power law is $\tau = 2.02 \pm 0.03$, which is in good agreement with the figure for the same quantity drawn from fossil data (see section 1.2.2.1).

with $F_i < \eta$ are wiped out and replaced by new species with randomly chosen F_i and B_i.

The net result is that species with low fitness are rapidly removed from the system. However, when a large coevolutionary avalanche takes place, many species receive new, randomly chosen fitness values, some of which will be low, and this process provides a "source" of low-fitness species for extinction events.

Interestingly, the distribution of extinction events in this model follows a power law, apparently regardless of the distribution from which the stress levels η are chosen (fig. 3.6). Roberts and Newman (1996) offered an analytical explanation of this result within a "mean field" framework similar to the one used by Flyvbjerg et al. (1993) for the original Bak–Sneppen model. However, what is particularly intriguing is that, even though the distribution of avalanche sizes in the model still possesses an exponent in the region of $3/2$ or less, the *extinction* distribution is steeper, with a measured exponent of $\tau = 2.02 \pm 0.03$ in excellent agreement with the results derived from the fossil data.

The model however has some disadvantages. First, the source of the power-law in the extinction distribution is almost certainly not a critical process, even though the Bak–Sneppen model, from which this model is derived, is critical. In fact, the model of Newman and Roberts is just a special case of the extinction model proposed later by Newman (see section 5.1), which does not contain any coevolutionary avalanches at all. In other words, the interesting behavior of the extinction distribution in this model is entirely independent of the coevolutionary behavior inherited from the Bak–Sneppen model.

A more serious problem with the model is the way in which the environmental stress is imposed. As we pointed out in section 3.1, the time steps in the Bak–Sneppen model correspond to different durations of geological time. This means that there should be a greater chance of a large stress hitting during time steps which correspond to longer periods. In the model of Newman and Roberts however, this is not the case; the probability of generating a given level of stress is the same in every time step. In the model of stress-driven extinction discussed in section 5.1 this shortcoming is rectified.

Another, very similar extension of the Bak–Sneppen model was introduced by Schmoltzi and Schuster (1995). Their motivation was somewhat different from that of Newman and Roberts—they were interested in introducing a "real time scale" into the model. As they put it: "The [Bak–Sneppen] model does not describe evolution on a physical time scale, because an update step *always* corresponds to a mutation of the species with the smallest fitness and its neighbors. This implies that we would observe constant extinction intensity in morphological data and that there will never be periods in which the system does not change." This is in fact only true if one ignores the rescaling of time implied by eq. (3.1). As figure 3.3 shows, there are very clear periods in which the system does not change if one calculates the time in the way Bak and Sneppen did.

The model of Schmoltzi and Schuster also incorporates an external stress term, but in their case it is a local stress η_i, varying from species to species. Other than that however, their approach is very similar to that of Newman and Roberts; species with fitness below η_i are removed from the system and replaced with new species, and all the variables $\{\eta_i\}$ are chosen anew at each time step. Their results also are rather similar to those of Newman and Roberts, although their main interest was to model neuronal dynamics in the brain, rather than evolution, so that they concentrated on somewhat different measurements. There is no mention of extinction, or of avalanche sizes, in their paper.

CHAPTER 4

Interspecies Connection Models

In the Bak–Sneppen model studied in the previous chapter there is no explicit notion of an interaction strength between two different species. It is true that if two species are closer together on the lattice, then there is a higher chance of their participating in the same avalanche. But beyond this there is no variation in the magnitude of the influence of one species on another. Real ecosystems, on the other hand, have a wide range of possible interactions between species, and as a result the extinction of one species can have a wide variety of effects on other species. These effects may be helpful or harmful, as well as strong or weak, and there is in general no symmetry between the effect of A on B and B on A. For example, if species A is prey for species B, then A's demise would make B less able to survive, perhaps driving it also to extinction, whereas B's demise would aid A's survival. On the other hand, if A and B compete for a common resource, then either's extinction would help the other. Or if A and B are in a mutually supportive or symbiotic relationship, then each would be hurt by the other's removal.

A number of authors have constructed models involving specific species–species interactions, or "connections." If species i depends on species j, then the extinction of j may also lead to the extinction of i, and possibly give rise to cascading avalanches of extinction. Most of these connection models neither introduce nor have need of a fitness measure, barrier, viability, or tolerance for the survival of individual species; the extinction pressure on one species comes from

53

the extinction of other species. Such a system still needs some underlying driving force to keep its dynamics from stagnating, but this can be introduced by making changes to the connections in the model, without requiring the introduction of any extra parameters.

Since the interactions in these models are ecological in nature (taking place at the individual level) rather than evolutionary (taking place at the species level or the level of the fitness landscape), the characteristic time scale of the dynamics is quite short. Extinctions produced by ecological effects such as predation and invasion can take only a single season, whereas those produced by evolutionary pressures are assumed to take much longer, maybe thousands of years or more.

The models described in this chapter vary principally in their connection topology, and in their mechanisms for replacing extinct species. Solé and co-workers have studied models with no organized topology, each species interacting with all others, or with a more-or-less random subset of them (Solé and Manrubia 1996; Solé, Bascompte and Manrubia 1996; Solé 1996). By contrast, the models of Amaral and Meyer (1999) and Abramson (1997) involve very specific food-chain topologies. The models of Solé et al. keep a fixed total number of species, refilling empty niches by invasion of surviving species. Abramson's model also keeps the total fixed, but fills empty niches with random new species, while Amaral and Meyer use an invasion mechanism, but do not attempt to keep the total number of species fixed.

4.1   THE SOLÉ–MANRUBIA MODEL

Solé and Manrubia (1996, Solé, Bascompte, and Manrubia 1996; Solé 1996) have constructed a model that focuses on species–species interactions through a "connection matrix" $\mathbf{J}$ whose elements give the strength of coupling between each pair of species. Specifically, the matrix element J_{ij} measures the influence of species i on species j, and J_{ji} that of j on i. A positive value of J_{ij} implies that i's continued existence helps j's survival, whereas a negative value implies that j would be happy to see i disappear. The J_{ij} values range between -1 and 1, chosen initially at random. In most of their work, Solé and Manrubia let every species interact with every other species, so all J_{ij}s are nonzero, though some may be quite small. Alternatively it is possible to define models in which the connections are more restricted, for instance by placing all the species on a square lattice and permitting each to interact only with its four neighbors (Solé 1996).

A species i becomes extinct if its net support $\sum_j J_{ji}$ from others drops below a certain threshold θ. The sum over j here is of course only over those species that (a) are not extinct themselves, and (b) interact with i (in the case of restricted connections). Solé and Manrubia introduce a variable $S_i(t)$ to represent whether species i is alive ($S_i = 1$) or extinct ($S_i = 0$) at time t, so the extinction dynamics may be written

$$S_i(t + 1) = \Theta\left[\sum_j J_{ji}S_j(t) - \theta\right],\qquad(4.1)$$

where $\Theta(x)$ is the Heaviside step function, which is 1 for $x > 0$ and zero otherwise. As this equation implies, time progresses in discrete steps, with all updates occurring simultaneously at each step. When avalanches of causally connected extinctions occur, they are necessarily spread over a sequence of successive time steps.

To complete the model, Solé and Manrubia introduce two further features, one to drive the system and one to replace extinct species. The driving force is simply a slow random mutation of the coupling strengths in the connection matrix $\mathbf{J}$. At each time step, for each species i, one of the incoming connections J_{ji} is chosen at random and given a new random value in the interval between -1 and 1. This may cause one or more species to become extinct though loss of positive support from other species or through increase in the negative influences on it. It is not essential to think of these mutations as strictly biotic; external environmental changes could also cause changes in the coupling between species (and hence in species' viability).

The replacement of extinct species is another distinguishing feature of Solé and Manrubia's model. All the niches that are left empty by extinction are immediately refilled with copies of one of the surviving species, chosen at random. This is similar to the speciation processes studied by Kauffman and Neumann in the variation of the NKCS model described in section 2.5, and in fact Solé and Manrubia refer to it as "speciation." However, because the Solé–Manrubia model is a model of ecological rather than evolutionary processes, it is probably better to think of the repopulation processes as being an invasion of empty niches by survivor species, rather than a speciation event. Speciation is inherently an evolutionary process, and, as discussed above, takes place on longer time scales than the ecological effects which are the primary concern of this model.

Invading species are copied to the empty slots along with all their incoming and outgoing connections, except that a little noise is added to these connections to introduce diversity. Specifically, if species k is copied to fill a number of open niches i, then

$$J_{ij} = J_{kj} + \eta_{ij}, \qquad J_{ji} = J_{jk} + \eta_{ji},\qquad(4.2)$$

where j ranges over the species with which each i interacts, and the ηs are all chosen independently from a uniform random distribution in the interval $(-\epsilon, \epsilon)$.

Because empty niches are immediately refilled, the $S_i(t)$ variables introduced on the right-hand side of eq. (4.1) are actually always 1, and are therefore superfluous. They do however make the form of the dynamics formally very similar to that of spin glasses in physics (Fischer and Hertz 1991), and to that of Hopfield artificial neural networks (Hertz et al. 1991), and it is possible that these similarities will lead to useful cross-fertilization between these areas of study.

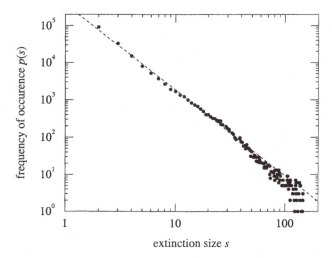

FIGURE 4.1   The distribution of sizes of extinction events in a simulation of the model of Solé and Manrubia (1996) with $N = 150$ species. The distribution follows a power law with a measured exponent of $\tau = 2.3 \pm 0.1$.

Solé and Manrubia studied their model by simulation, generally using $N = 100$ to 150 species, $\theta = 0$, and $\epsilon = 0.01$. Starting from an initial random state, they waited about 10,000 time steps for transients to die down before taking data. Extinction events in the model were found to range widely in size s, including occasional large "mass extinction" events that wiped out over 90% of the population. Such large events were often followed by a long period with very little activity. The distribution $p(s)$ of extinction sizes was found to follow a power law, as in eq. (1.1), with $\tau = 2.3 \pm 0.1$ (see fig. 4.1). Later work by Solé et al. (1996) using $\epsilon = 0.05$ gave $\tau = 2.05 \pm 0.06$, consistent with the value $\tau = 2.0 \pm 0.2$ from the fossil data (section 1.2.2.1).

The diversified descendants of a parent species may be thought of as a single genus, all sharing a common ancestor. Since the number of offspring of a parent species is proportional to the number of niches which need to be filled following a extinction event, the distribution of genus sizes is exactly the same as that of extinction sizes. Thus Solé and Manrubia find an exponent in the vicinity of 2 for the taxonomic distribution as well (see eq. (1.2)), to be compared to 1.5 ± 0.1 for Willis's data (fig. 1.11) and to values between 1.1 and 2.1 for Burlando's analysis (section 3.5).

The waiting time between two successive extinction events in the Solé–Manrubia model is also found to have a power-law distribution, with exponent 3.0 ± 0.1. Thus events are correlated in time—a random (Poisson) process would have an exponential distribution of waiting times. The distribution of both species and genus lifetimes can in theory also be measured in these simulations, although Solé and Manrubia did not publish any results for these quantities. Further studies would be helpful here.

Solé and Manrubia claim on the basis of their observed power laws that their model is self-organized critical. However, it turns out that this is not the case (Solé, private communication). In fact, the model is an example of an ordinary critical system which is tuned to criticality by varying the parameter θ, which is the threshold at which species become extinct. It is just coincidence that the value $\theta = 0$ which Solé and Manrubia used in all of their simulations is precisely the critical value of the model at which power laws are generated. Away from this value the distributions of the sizes of extinction events and of waiting times are cut off exponentially at some finite scale, and therefore do not follow a power law. This then begs the question of whether there is any reason why in a real ecosystem this threshold parameter should take precisely the value which produces the power law distribution, rather than any other value. At present, no persuasive case has been made in favor of $\theta = 0$, and so the question remains open.

4.2   VARIATIONS ON THE SOLÉ–MANRUBIA MODEL

A number of variations on the basic model of Solé and Manrubia are mentioned briefly in the original paper (Solé and Manrubia 1996). The authors tried relaxing the assumptions of total connectivity (letting some pairs of species have no influence on each other), of $\theta = 0$, and of diversification (letting $\epsilon = 0$). They also tried letting each J_{ij} take only the values $+1$ or -1. In all these cases they report that they found the same behavior with the same power-law exponents (although as mentioned above, later results showed that in fact the power-law behavior is destroyed by making $\theta \neq 0$). This robustness to changing assumptions is to be expected for critical phenomena, where typically there occur large "universality classes" of similar behavior with identical exponents (see section 3.2).

Solé (1996) presents a more significant extension of the model which does change some of the exponents: he proposes a dynamical rule for the connectivity itself. At any time some pairs of sites i, j are not connected, so that in effect $J_{ij} = J_{ji} = 0$. (Solé introduces a new connection variable to represent this, but that is not strictly necessary.) Initially the number of connections per site is chosen randomly between 1 and $N - 1$. During the population of an empty niche i by a species k, all but one of k's nonzero connections are reproduced with noise, as in eq. (4.2), but the last is discarded and replaced entirely by a new random link from i to a site to which k is *not* connected.

Solé also replaces the mutation of J_{ij}, which provides the fundamental random driving force in the Solé–Manrubia model, by a rule that removes one of the existing species at random at any step when no extinction takes place. Without this driving force the system would in general become frozen. The emptied niche is refilled by invasion as always, but these "random" extinction events are not counted in the statistical analysis of extinction. (The waiting time would always be 1 if they were counted.) It is not clear whether this difference between the models has a significant effect on the results.

The observed behavior of this model is similar to that of the Solé–Manrubia model as far as extinction sizes are concerned; Solé reports an exponent $\tau = 2.02 \pm 0.03$ for the extinction size distribution. However the waiting-time distribution falls much more slowly (so there are comparably more long waits), with an exponent 1.35 ± 0.07 compared to 3.0 ± 0.1 for the Solé–Manrubia model. The smaller exponent seems more reasonable, though of course experimental waiting time data is not available for comparison. The number of connections itself varies randomly through time, and a Fourier analysis shows a power spectrum of the form $1/f^{\nu}$ with $\nu = 0.99 \pm 0.08$. Power spectra of this type are another common feature of critical systems (Solé et al. 1997).

4.3   AMARAL AND MEYER'S FOOD CHAIN MODEL

Whereas the Solé–Manrubia model and its variants have a more or less arbitrary connection topology between species, real ecosystems have very specific sets of interdependencies. An important part of the natural case can be expressed in terms of food chains, specifying who eats whom. Of course food chains are not the only type of interspecies interaction, but it is nevertheless of interest to consider models of extinction based on food-chain dynamics. Amaral and Meyer (1999) and Abramson (1997) have both constructed and studied such models.

Amaral and Meyer (1999) have proposed a model in which species are arranged in L trophic levels labeled $l = 0, 1, \ldots, L - 1$. Each level has N niches, each of which may be occupied or unoccupied by a species. A species in level l (except $l = 0$) feeds on up to k species in level $l - 1$; these are its prey. If all of a species' prey become extinct, then it too becomes extinct, so avalanches of extinction can occur. This process is driven by randomly selecting one species at level 0 at each time step and making it extinct with probability p. There is no sense of fitness or of competition between species governing extinction in this model.

To replace extinct species, Amaral and Meyer use a speciation mechanism. At a rate μ, each existing species tries to engender an offspring species by picking a niche at random in its own level or in the level above or below. If that randomly selected niche is unoccupied, then the new species is created and assigned k preys at random from the existing species on the level below. The parameter μ needs to be large enough that the average origination rate exceeds the extinction rate, or all species will become extinct. Note that, as pointed out earlier, speciation is inherently an evolutionary process and typically takes place on longer time scales than extinction through ecological interactions, so there is some question about whether it is appropriate in a model such as this. As with the Solé–Manrubia model, it might be preferable to view the repopulation of niches as an invasion process, rather than a speciation one.

The model is initialized by populating the first level $l = 0$ with some number N_0 of species at random. Assuming a large enough origination rate, the population will then grow approximately exponentially until limited by the number of available niches.

Amaral and Meyer presented results for a simulation of their model with parameters $L = 6$, $k = 3$, $N = 1000$, $N_0 \approx 50$, $p = 0.01$, and $\mu = 0.02$. The statistics of extinction events are similar to those seen in many other models. The time series is highly intermittent, with occasional large extinction events almost up to the maximum possible size NL. The distribution of extinction sizes s fits a power law, eq. (1.1), with exponent $\tau = 1.97 \pm 0.05$. Origination rates are also highly intermittent, and strongly correlated with extinction events.[10]

An advantage of this model is that the number of species is not fixed, and its fluctuations can be studied and compared with empirical data. Amaral and Meyer compute a power spectrum for the number of species and find that it fits a power law $p(f) \propto 1/f^\nu$ with $\nu = 1.95 \pm 0.05$. The authors argue that this reveals a "fractal structure" in the data, but it is worth noting that a power-spectrum exponent of $\nu = 2$ occurs for many nonfractal processes, such as simple random walks, and a self-similar structure only needs to be invoked if $\nu < 2$.

Amaral and Meyer also compute a power spectrum for the extinction rate, for comparison with the fossil data analysis of Solé et al. (1997). They find a power law with $\nu \simeq 1$ for short sequences, but then see a crossover to $\nu \simeq 0$ at longer times, suggesting that there is no long-time correlation.

Drossel (1999) has analyzed the Amaral–Meyer model in some detail. The $k = 1$ case is most amenable to analysis, because then the food chains are simple independent trees, each rooted in a single species at level 0. The extinction size distribution is therefore equal to the tree size distribution, which can be computed by master equation methods, leading to $p(s) \propto s^{-2}$ (i.e., $\tau = 2$) exactly in the limits $N \to \infty$, $L \to \infty$. Finite size effects (when N or L are not infinite) can also be evaluated, leading to a cutoff in the power law at $s_{\max} \sim N \log N$ if $L \gg \log N$ or $s_{\max} \sim e^L$ if $L \ll \log N$. These analytical results agree well with the simulation studies.

The analysis for $k > 1$ is harder, but can be reduced in the case of large enough L and N (with $L \ll \ln N$) to a recursion relation connecting the lifetime distribution of species on successive levels. This leads to the conclusion that the lifetime distribution becomes invariant after the first few levels, which in turn allows for a solution. The result is again a power-law extinction size distribution with $\tau = 2$ and cutoff $s_{\max} \sim e^L$.

Drossel also considers a variant of the Amaral–Meyer model in which a species becomes extinct if *any* (instead of all) of its prey disappear. She shows

---

[10]The authors report that they obtained similar results, with the same exponents, for larger values of k too (Amaral, private communication).

that this too leads to a power law with $\tau = 2$, although very large system sizes would be needed to make this observable in simulation. She also points out that other variations of the model (such as making the speciation rate depend on the density of species in a layer) do not give power laws at all, so one must be careful about attributing too much universality to the "critical" nature of this model.

4.4   ABRAMSON'S FOOD CHAIN MODEL

Abramson (1997) has proposed a different food chain model in which each species is explicitly represented as a population of individuals. In this way Abramson's model connects extinction to microevolution, rather than being a purely macroevolutionary model. There is not yet a consensus on whether a theory of macroevolution can be built solely on microevolutionary principles; see Stenseth (1985) for a review.

Abramson considers only linear food chains, in which a series of species at levels $i = 1, 2, \ldots, N$ each feed on the one below (except $i = 1$) and are fed on by the one above (except $i = N$). If the population density at level i at time t is designated by $n_i(t)$, then the changes in one time step are given by

$$n_i(t+1) - n_i(t) = k_i n_{i-1}(t) n_i(t)[1 - n_i(t)/c_i]$$
$$-g_i n_{i+1}(t) n_i(t). \tag{4.3}$$

Here k_i and g_i represent the predation and prey rates, and c_i is the carrying capacity of level i. These equations are typical of population ecology. At the endpoints of the chain, boundary conditions may be imposed by adjoining two fictitious species, 0 and $N + 1$ with $n_0 = n_{N+1} = 1$. For simplicity Abramson takes $c_i = 1$ for all i, and sets $g_i = k_{i+1}$. The species are then parametrized simply by their k_i and by their population size $n_i(t)$. These are initially chosen randomly in the interval $(0, 1)$.

The population dynamics typically leads to some $n_i(t)$'s dropping asymptotically to 0. When they drop below a small threshold, Abramson regards that species as extinct and replaces it with a new species with randomly chosen n_i and k_i, drawn from uniform distributions in the interval between zero and one. But an additional driving force is still needed to prevent the dynamics from stagnating. So with probability p at each time step, Abramson also replaces one randomly chosen species, as if it had become extinct.

The replacement of an extinct species by a new one with a larger population size has in general a negative impact on the species below it in the food chain. Thus one extinction event can lead to an avalanche propagating down the food chain. Note that this is the precise opposite of the avalanches in the Amaral–Meyer model, which propagate upwards due to loss of food sources.

Abramson studies the statistics of extinction events in simulations of his model for values of N from 50 to 1000.[11] He finds punctuated equilibrium in the extinction event sizes, but the size distribution $p(s)$ does *not* fit a power law. It does show some scaling behavior with N, namely $p(s) = N^\beta f(sN^\nu)$, where β and ν are parameters and $f(x)$ is a particular "scaling function." Abramson attributes this form to the system being in a "critical state." The waiting time between successive extinctions fits a power law over several decades of time, but the exponent seems to vary with the system size. Overall, this model does not have strong claims for criticality and does not agree very well with the extinction data.

---

[11]Solé (private communication) has made the point that these values are unrealistically large for real food chains. Real food chains typically have less than ten trophic levels.

Environmental Stress Models

In chapters 2 to 4 we discussed several models of extinction which make use of ideas drawn from the study of critical phenomena. The primary impetus for this approach was the observation of apparent power-law distributions in a variety of statistics drawn from the fossil record, as discussed in section 1.2; in other branches of science such power laws are often indicators of critical processes. However, there are also a number of other mechanisms by which power laws can arise, including random multiplicative processes (Montroll and Shlesinger 1982; Sornette and Cont 1997), extremal random processes (Sibani and Littlewood 1993), and random barrier-crossing dynamics (Sneppen 1995). Thus the existence of power-law distributions in the fossil data is not on its own sufficient to demonstrate the presence of critical phenomena in extinction processes.

Critical models also assume that extinction is caused primarily by biotic effects such as competition and predation, an assumption which is in disagreement with the fossil record. As discussed in section 1.2.2.1, all the plausible causes for specific prehistoric extinctions are abiotic in nature. Therefore an obvious question to ask is whether it is possible to construct models in which extinction is caused by abiotic environmental factors, rather than by critical fluctuations arising out of biotic interactions, but which still give power-law distributions of the relevant quantities.

Such models have been suggested by Newman (1996, 1997) and by Manrubia and Paczuski (1998). Interestingly, both of these models are the result of attempts at simplifying models based on critical phenomena. Newman's model is a simplification of the model of Newman and Roberts (see section 3.6), which included both biotic and abiotic effects; the simplification arises from the realization that the biotic part can be omitted without losing the power-law distributions. Manrubia and Paczuski's model was a simplification of the connection model of Solé and Manrubia (see section 4.1), but in fact all direct species-species interactions were dropped, leaving a model which one can regard as driven only by abiotic effects. We discuss these models in turn.

5.1 NEWMAN'S STRESS MODEL

The model proposed by Newman (1996, 1997) has a fixed number N of species which in the simplest case are noninteracting. Real species do interact of course, but as we will see the predictions of the model are not greatly changed if one introduces interactions, and the noninteracting version makes a good starting point because of its extreme simplicity. The absence of interactions between species also means that critical fluctuations cannot arise, so any power laws produced by the model are definitely of noncritical origin.

As in the model of Newman and Roberts (1995), the level of the environmental stress is represented by a single number η, which is chosen independently at random from some distribution $p_{\text{stress}}(\eta)$ at each time step. Each species $i = 1 \ldots N$ possesses some threshold tolerance for stress denoted x_i which is high in species which are well able to withstand stress and low in those which are not. (See Jablonski [1989] for a discussion of the selectivity of extinction events in the fossil record.) Extinction takes place via a simple rule: if at any time step the numerical value of the stress level exceeds a species' tolerance for stress, $\eta > x_i$, then that species becomes extinct at that time step. Thus large stresses (sea-level change, bolide impact) can give rise to large mass extinction events, while lower levels of stress produce less dramatic background extinctions. Note that simultaneous extinction of many species occurs in this model because the same large stress affects all species, and not because of any avalanche or domino effects in the ecosystem.

In order to maintain a constant number of species, the system is repopulated after every time step with as many new species as have just become extinct. The extinction thresholds x_i for the new species can either be inherited from surviving species, or can be chosen at random from some distribution $p_{\text{thresh}}(x)$. To a large extent it appears that the predictions of the model do not depend on which choice is made; here we focus on the uniform case with $p_{\text{thresh}}(x)$ a constant independent of x over some allowed range of x, usually $0 \leq x < 1$. In addition, it is safe to assume that the initial values of the variables x_i are also chosen according to $p_{\text{thresh}}(x)$, since in any case the effects of the initial choices only persist as long as it takes to turn over all the species in the ecosystem, which happens many times during a run of the model (and indeed many times during the known fossil record).

There is one further element which needs to be added to the model in order to make it work. As described, the species in the system start off with randomly chosen tolerances x_i and, through the extinction mechanism described above, those with the lowest tolerance are systematically removed from the population and replaced by new species. Thus, the number of species with low thresholds for extinction decreases over time, in effect creating a gap in the distribution, as in the Bak–Sneppen model. As a result the size of the extinction events taking place dwindles and ultimately extinction ceases almost entirely, a behavior which we know not to be representative of a real ecosystem. Newman suggests that the solution to this problem comes from evolution. In the intervals between large stress events, species will evolve under other selection pressures, and this will change the values of the variables x_i in unpredictable ways. Adapting to any particular selection pressure might raise, lower, or leave unchanged a species' tolerance to environmental stresses. Mathematically this is represented by making random changes to the x_i, either by changing them all slightly at each time step, or by changing a small fraction f of them to totally new values drawn from $p_{thresh}(x)$, and leaving the rest unchanged. These two approaches can be thought of as corresponding to gradualist and punctuationalist views of evolution respectively, but it appears in practice that the model's predictions are largely independent of which is chosen. In his work Newman focused on the punctuationalist approach, replacing a fraction f of the species by random new values.

This description fully defines Newman's model except for the specification of $p_{stress}(\eta)$ and $p_{thresh}(x)$. However, it turns out that we can, without loss of generality, choose $p_{thresh}(x)$ to have the simple form of a uniform distribution in the interval from 0 to 1, since any other choice can be mapped onto this with the transformation

$$x \to x' = \int_{-\infty}^{x} p_{thresh}(y) \, dy. \tag{5.1}$$

The stress level must of course be transformed in the same way, $\eta \to \eta'$, so that the condition $\eta' > x_i'$ corresponds precisely to $\eta > x_i$. This in turn requires a transformation

$$p_{stress}(\eta') = p_{stress}(\eta)\frac{d\eta}{d\eta'} = \frac{p_{stress}(\eta)}{p_{thresh}(\eta)} \tag{5.2}$$

for the stress distribution.

The choice of $p_{stress}(\eta)$ remains a problem, since it is not known what the appropriate distribution of stresses is in the real world. For some particular sources of stress, such as meteor impacts, there are reasonably good experimental results for the distribution (Morrison 1992; Grieve and Shoemaker 1994), but overall we have very little knowledge about stresses occurring either today or in the geologic past. Newman, therefore, tested the model with a wide variety of

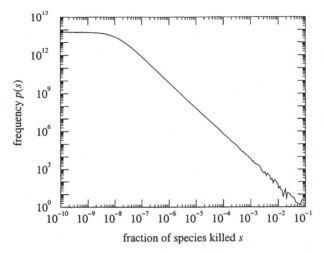

FIGURE 5.1   Distribution of the sizes of extinction events taking place in the model of Newman (1996). The distribution is power-law in form with an exponent of $\tau = 2.02 \pm 0.02$ except for extinctions of very small size, where it becomes flat.

stress distributions and found that, in a fashion reminiscent of the self-organized critical models, many of its predictions are robust against variations in the form of $p_{\text{stress}}(\eta)$, within certain limits.

In figure 5.1 we show simulation results for the distribution $p(s)$ of the sizes s of extinction events in the model for one particular choice of stress distribution, the Gaussian distribution:

$$p_{\text{stress}}(\eta) \propto \exp\left[-\frac{\eta^2}{2\sigma^2}\right]. \tag{5.3}$$

This is probably the most common noise distribution occurring in natural phenomena. It arises as a result of the central limit theorem whenever a number of different independent random effects combine additively to give one overall stress level. As the figure shows, the resulting distribution of the sizes of extinction events in Newman's model follows a power law closely over many decades. The exponent of the power law is measured to be $\tau = 2.02 \pm 0.02$, which is in good agreement with the value of 2.0 ± 0.2 found in the fossil data. The only deviation from the power-law form is for very small sizes s, in this case below about one species in 10^8, where the distribution flattens off and becomes independent of s. The point at which this happens is controlled primarily by the value of the parameter f, which governs the rate of evolution of species (Newman and Sneppen 1996). No flat region is visible in the fossil extinction distribution, figure 1.1, which implies that the value of f must be small—smaller than the smallest fractional extinction which can be observed reliably in fossil data. How-

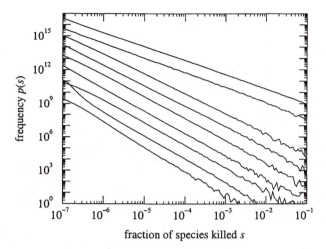

FIGURE 5.2 Distribution of the sizes of extinction events for a variety of different stress distributions, including Gaussian, Lorentzian, Poissonian, exponential, and
stretched exponential. In each case the distribution follows a power law closely over
many decades.

ever, this is not a very stringent condition, since it is not possible to measure
extinctions smaller than a few percent with any certainty.

In figure 5.2 we show results for the extinction size distribution for a wide
variety of other distributions $p_{\text{stress}}(\eta)$ of the applied stress, including various
different Gaussian forms, exponential and Poissonian noise, power laws, and
stretched exponentials. As the figure shows, the distribution takes a power-law
form in each case. The exponent of the power law varies slightly from one curve
to another, but in all cases it is fairly close to the value of $\tau \simeq 2$ found in the
fossil record. In fact, Sneppen and Newman (1997) have shown analytically that
for all stress distributions $p_{\text{stress}}(\eta)$ satisfying

$$\int_{\eta}^{\infty} p_{\text{stress}}(x)\, \mathrm{d}x \sim p_{\text{stress}}(\eta)^{\alpha} \qquad (5.4)$$

for large η and some exponent α, the distribution of extinction sizes will take
a power-law form for large s. This condition is exactly true for exponential and
power-law distributions of stress, and approximately true for Gaussian and Poissonian distributions. Since this list covers almost all noise distributions which
occur commonly in natural systems, the predictions of the model should be reasonably robust, regardless of the ultimate source of the stresses.

It is also straightforward to measure the lifetimes of species in simulations
of this model. Figure 5.3 shows the distribution of lifetimes measured in one
particular run. The distribution is power-law in form as it is in the fossil data,
with a measured exponent of 1.03 ± 0.05.

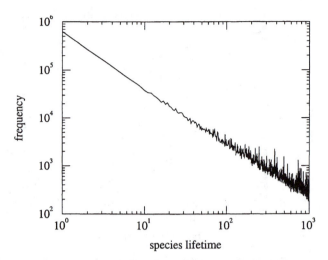

FIGURE 5.3   The distribution of the lifetimes of species in the model of Newman (1997). The distribution follows a power law with an exponent in the vicinity
of 1.

Newman (1997) has given a number of other predictions of his model. In
particular, he has suggested how taxonomy can be incorporated into the model
to allow one to study the birth and death of genera and higher taxa, in addition to
species. With this extension the model predicts a distribution of genus lifetimes
similar to that of species, with a power-law form and exponent in the vicinity
of one. Note that although the power-law form is seen also in the fossil data, an
exponent of one is not in agreement with the value of 1.7 ± 0.3 measured in the
fossil lifetime distribution (see section 1.2.2.4). The model does however correctly
predict Willis's power-law distribution of the number of species per genus (see
section 1.2.3.1) with an exponent close to the measured value of $\beta = 3/2$.

Another interesting prediction of the model is that of "aftershock extinctions"—strings of smaller extinctions arising in the aftermath of a large mass
extinction event (Sneppen and Newman 1997; Wilke et al. 1998). The mechanism behind these aftershock extinctions is that the repopulation of ecospace
after a large event tends to introduce an unusually high number of species with
low tolerance for stress. (At other times such species are rarely present because
they are removed by the frequent small stresses applied to the system.) The rapid
extinction of these unfit species produces a high turnover of species for a short
period after a mass extinction, which we see as a series of smaller "aftershocks."
The model makes the particular prediction that the intervals between these aftershock extinctions should fall off with time as t^{-1} following the initial large
event. This behavior is quite different from that of the critical models of earlier chapters, and therefore it could provide a way of distinguishing in the fossil
record between the two processes represented by these models. So far, however,

no serious effort has been made to look for aftershock extinctions in the fossil data, and indeed it is not even clear that the available data are adequate for the task. In addition, later work by Wilke and Martinetz (1997) calls into question whether one can expect aftershocks to occur in real ecosystems. (This point is discussed further in section 5.4.)

5.2 SHORTCOMINGS OF THE MODEL

Although Newman's model is simple and makes predictions which are in many cases in good agreement with the fossil data, there are a number of problems associated with it.

First, one could criticize the assumptions which go into the model. For example, the model assumes that species are entirely noninteracting, which is clearly false. In the version we have described here it also assumes a "punctuated" view of evolution in which species remain constant for long periods and then change abruptly. In addition, the way in which new species are added to the model is questionable: new species are given a tolerance x_i for stress which is chosen purely at random, whereas in reality new species are presumably descended from other earlier species and therefore one might expect some correlation between the values of x_i for a species and its ancestors.

These criticisms lead to a number of generalizations of the model which have been examined by Newman (1997). To investigate the effect of species interactions, Newman looked at a variation of the model in which the extinction of a species could give rise to the extinction of a neighboring species, in a way reminiscent of the avalanches of Kauffman's NK model. He placed the model on a lattice and added a step to the dynamics in which the extinction of a species as a result of external stress caused the knock-on extinction (and subsequent replacement) of all the species on adjacent lattice sites. In simulations of this version of the model, Newman found, inevitably, spatial correlations between the species becoming extinct which are not present in the original version. Other than this however, it appears that the model's predictions are largely unchanged. The distributions of extinction event sizes and taxon lifetimes for example are still power-law in form and still possess approximately the same exponents.

Similarly it is possible to construct a version of the model in which evolution proceeds in a "gradualist" fashion, with the values of the variables x_i performing a slow random walk rather than making punctuated jumps to unrelated values. And one can also create a version in which the values of x_i assumed by newly appearing species are inherited from survivors, rather than chosen completely at random. Again it appears that these changes have little effect on the major predictions of the model, although these results come primarily from simulations of the model; the analytic results for the simplest version do not extend to the more sophisticated models discussed here.

5.3   THE MULTITRAIT VERSION OF THE MODEL

A more serious criticism of Newman's model is that it models different types of stress using only a single parameter η. Within this model one can only say whether the stress level is high or low at a particular time. In the real world there are many different kinds of stress, such as climatic stress, ecological stresses like competition and predation, disease, bolide impact, changes in ocean chemistry, and many more. And there is no guarantee that a period when one type of stress is high will necessarily correspond to high stress of another type. This clearly has an impact on extinction profiles, since some species will be more susceptible to stresses of a certain kind than others. To give an example, it is thought that large body mass was a contributing factor to extinction at the Cretaceous–Tertiary boundary (Clemens 1986). Thus the particular stress which caused the K–T extinction, thought to be the result of a meteor impact, should correspond to tolerance variables x_i in our model which are lower for large-bodied animals. Another type of stress—sea-level change, say—may have little or no correlation with body size.

To address this problem, Newman (1997) has also looked at a variation of his model in which there is a number M of different kinds of stress. In this case each species also has a separate tolerance variable $x_i^{(k)}$ for each type of stress k and becomes extinct if any one of the stress levels exceeds the corresponding threshold. As with the other variations on the model, it appears that this "multitrait" version reproduces the important features of the simpler versions, including the power-law distributions of the sizes of extinction events and of species lifetimes. Sneppen and Newman (1997) have explained this result with the following argument. To a first approximation, one can treat the probability of a species becoming extinct in the multitrait model as the probability that the stress level exceeds the lowest of the thresholds for stress which that species possesses. In this case, the multitrait model is identical to the single-trait version but with a different choice for the distribution $p_{\text{thresh}}(x)$ from which the thresholds are drawn (one which reflects the probability distribution of the lowest of M random numbers). However, as we argued earlier, the behavior of the model is independent of $p_{\text{thresh}}(x)$ since we can map any distribution on the uniform one by a simple integral transformation of x (see eq. (5.1)).

5.4   THE FINITE-GROWTH VERSION OF THE MODEL

Another shortcoming of the model proposed by Newman is that the species which become extinct are replaced instantly by an equal number of new species. In reality, fossil data indicate that the process of replacement of species takes a significant amount of time, sometimes as much as a few million years (Stanley 1990; Erwin 1996). Wilke and Martinetz (1997) have proposed a generalization of the model which takes this into account. In this version, species which become extinct are replaced slowly according to the logistic growth law

$$\frac{\mathrm{d}N}{\mathrm{d}t} = gN(1 - N/N_{\mathrm{max}}),\tag{5.5}$$

where N is the number of species as before, and g and N_{max} are constants. Logistic growth appears to be a reasonable model for recovery after large extinction events (Sepkoski 1991; Courtillot and Gaudemer 1996). When the growth parameter g is infinite, we recover the model proposed by Newman. Wilke and Martinetz find, as one might expect, that there is a transition in the behavior of the system at a critical value $g = g_c$ where the rate of repopulation of the system equals the average rate of extinction. They give an analytic treatment of the model which shows how g_c varies with the other parameters in the problem. For values of g below g_c, life eventually dies out in the model, and it is probably reasonable to assume that the Earth is not, for the moment at least, in this regime. For values of g above g_c it is found that the power-law behavior seen in the simplest versions of the model is retained. The value of the extinction size exponent τ appears to decrease slightly with increasing g, but is still in the vicinity of the value $\tau \simeq 2$ extracted from the fossil data. Interestingly they also find that the aftershock extinctions discussed in section 5.1 become less well defined for finite values of g, calling into question Newman's contention that the existence of aftershocks in the fossil record could be used as evidence in favor of his model. This point is discussed further by Wilke et al. (1998).

5.5 THE MODEL OF MANRUBIA AND PACZUSKI

Another variation on the ideas contained in Newman's model has been proposed by Manrubia and Paczuski (1998). Interestingly, although this model is mathematically similar to the other models discussed in this chapter, its inspiration is completely different. In fact, it was originally intended as a simplification of the connection model of Solé and Manrubia discussed in section 4.1.

In Newman's model, there are a large number of species with essentially constant fitness or tolerance to external stress, and those which fall below some time-varying threshold level become extinct. In the model of Manrubia and Paczuski by contrast, the threshold at which species become extinct is fixed and their fitness is varied over time. In detail, the model is as follows.

The model contains a fixed number N of species, each with a fitness x_i, or "viability" as Manrubia and Paczuski have called it. This viability measures how far a species is from becoming extinct, and might be thought of as a measure of reproductive success. All species are subject to random coherent stresses, or "shocks," which additively increase or decrease the viability of all species by the same amount η. If at any point the viability of a species falls below a certain threshold x_0, that species becomes extinct and is replaced by speciation from one of the surviving species. In Newman's model there was also an "evolution"

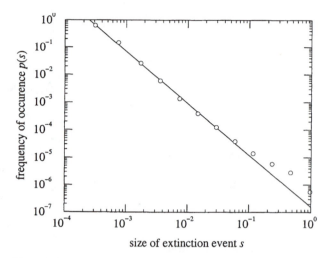

size of extinction event s

FIGURE 5.4   The distribution of the sizes of extinction events in a simulation of the model of Manrubia and Paczuski, with $N = 3200$ species (circles). The best-fit power law (solid line) has an exponent of $\tau = 1.88 \pm 0.09$. After Manrubia and Paczuski (1998).

process which caused species with high viability to drift to lower values over the course of time, preventing the system from stagnating when all species with low viability had been removed. The model of Manrubia and Paczuski contains an equivalent mechanism, whereby the viabilities of all species drift, in a stochastic fashion, toward lower values over the course of time. This also prevents stagnation of the dynamics.

Although no one has shown whether the model of Manrubia and Paczuski can be mapped exactly onto Newman's model, it is clear that the dynamics of the two are closely similar, and therefore it is not surprising to learn that the behavior of the two models is also similar. Figure 5.4 shows the distribution of the sizes s of extinction events in a simulation of the model with $N = 3200$ species. The distribution is close to power-law in form with an exponent of $\tau = 1.9$ similar to that of Newman's model, and in agreement with the result $\tau \simeq 2$ seen in the fossil data. The model also generates a power-law distribution in the lifetimes of species and, as in Newman's model, a simple definition of genus can be introduced and it can be shown that the distribution of number of species per genus follows a power law as well. The exponent of the lifetime distribution turns out to be approximately 2, which is not far from the value of 1.7 ± 0.3 found in the fossil data (see section 1.2.2.4).[12]

---

[12]The exponent for the distribution of genus sizes is also 2 which is perhaps a shortcoming of this model; recall that Willis's value for flowering plants was 1.5 (fig. 1.11), and the comprehensive studies by Burlando (1990, 1993) gave an average value of 1.6.

What is interesting about this model however, is that its dynamics is derived using a completely different argument from the one employed by Newman. The basic justification of the model goes like this. We assume first of all that it is possible to define a viability x_i for species i, which measures in some fashion how far a species is from the point of extinction. The point of extinction itself is represented by the threshold value x_0. The gradual downward drift of species' viability can be then be accounted for as the result of mutation; the majority of mutations lower the viability of the host.

Manrubia and Paczuski justify the coherent stresses in the system by analogy with the model of Solé and Manrubia (1996) in which species feel the ecological "shock" of the extinction of other nearby species. In the current model, the origin of the shocks is similarly taken to be the extinction of other species in the system. In other words it is the result of biotic interaction, rather than exogenous environmental influences. However, by representing these shocks as coherent effects which influence all species simultaneously to the same degree, Manrubia and Paczuski have removed from the dynamics the direct interaction between species which was present in the original connection model. Amongst other things, this allows them to give an approximate analytic treatment of their model using a time-averaged approximation similar to the one employed by Sneppen and Newman (1997) for Newman's model.

One further nice feature of the Manrubia–Paczuski model is that it is particularly easy in this case to see how large extinction events arise. Because species are replaced by speciation from others, the values of their viabilities tend to cluster together: most species are copies, or near copies, of other species in the system. Such clusters of species tend all to become extinct around the same time because they all feel the same coherent shocks and are all driven below the extinction threshold together. (A similar behavior is seen in the Solé–Manrubia model of section 4.1.) This clustering and avalanche behavior in the model is reminiscent of the so-called "phase coherent" models which have been proposed as a mechanism for the synchronization of the flashing of fireflies (Strogatz and Stewart 1993). Although no one has yet made a direct connection between these two classes of models, it is possible that mathematical techniques similar to those employed with phase-coherent models may prove profitable with models of type proposed by Manrubia and Paczuski.

CHAPTER 6

Non-Equilibrium Models

Sibani and co-workers have proposed a model of the extinction process, which they call the "reset model" (Sibani et al. 1995, 1998), which differs from those discussed in the preceding chapters in a fundamental way; it allows for, and indeed relies upon, nonstationarity in the extinction process. That is, it acknowledges that the extinction record is not uniform in time, is not in any sense in equilibrium, as it is assumed to be in the other models we have considered. In fact, extinction intensity has declined on average over time from the beginning of the Phanerozoic until the Recent. Within the model of Sibani et al., the distributions of section 1.2 are all the result of this decline, and the challenge is then to explain the decline, rather than the distributions themselves.

6.1 EXTINCTION RATE DECLINE

In figure 1.9 we showed the number of known families as a function of time over the last 600 My. On the logarithmic scale of the figure, this number appears to increase fairly steadily and although, as we pointed out, some of this increase can be accounted for by the bias known as the "pull of the recent," there is probably a real trend present as well. It is less clear that there is a similar trend in extinction intensity. The extinctions represented by the points in figure 1.1 certainly vary in intensity, but on average they appear fairly constant. Recall

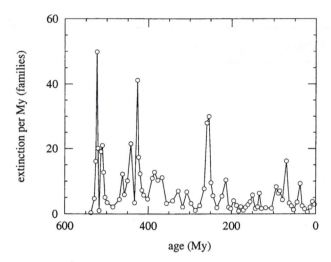

FIGURE 6.1 The number of families of marine organisms becoming extinct per million years in each of the stages of the Phanerozoic. The decline in average extinction rate is clearly visible in this plot. The data are from the compilation by Sepkoski (1992).

however, that figure 1.1 shows the number of families becoming extinct in each stage, and that the lengths of the stages are not uniform. In figure 6.1 we show the extinction intensity normalized by the lengths of the stages—the extinction rate in families per million years—and on this figure it is much clearer that there is an overall decline in extinction towards the Recent.

In order to quantify the decline in extinction rate, we consider the *cumulative* extinction intensity $c(t)$ as a function of time. The cumulative extinction at time t is defined to be the number of taxa which have become extinct up to that time. In other words, if we denote the extinction intensity at time t by $x(t)$, then the cumulative extinction intensity is

$$c(t) = \int_0^t x(t') \, dt' \,. \tag{6.1}$$

Figure 6.2 shows this quantity for the marine families in Sepkoski's database. Clearly the plot has to be monotonically increasing. Sibani et al. suggested that it in fact has a power-law form, with an exponent in the vicinity of 0.6. Newman and Eble (1999a) however have pointed out that it more closely follows a logarithmic increase law—a straight line on the linear–log scales of figure 6.2. (For comparison we show the same data on log–log scales in the inset. The power-law form proposed by Sibani et al. would appear as a straight line on these scales.) This implies that $c(t)$ can be written in the form

$$c(t) = A + B \log(t - t_0) \,, \tag{6.2}$$

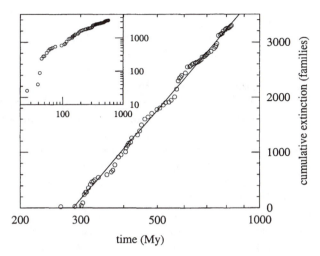

FIGURE 6.2 Main figure: the cumulative extinction intensity as a function of time during the Phanerozoic on linear–log scales. The straight line is the best logarithmic fit to the data. Inset: the same data on log–log scales. After Newman and Eble (1999a).

where A and B are constants and t_0 is the point of intercept of the line in figure 6.2 with the horizontal axis. (Note that t_0 lies before the beginning of the Cambrian. If time is measured from $t = 0$ at the start of the data set, which coincides roughly with the beginning of the Cambrian, then the best fit of the form (6.2) has $t_0 \simeq -260$ My.)

Combining eqs. (6.1) and (6.2) and differentiating with respect to t we get an expression for the extinction per unit time:

$$x(t) = \frac{B}{t - t_0} \,. \tag{6.3}$$

In other words the average extinction rate is falling off over time as a power law with exponent 1. Sibani et al. have pointed out that a power-law decline in itself could be enough to explain the distribution of the sizes of extinction events seen in figure 1.3. For an extinction profile of the form of eq. (6.3) the number of time intervals in which we expect to see extinction events of a certain size s is given by

$$p(s) = \left. \frac{\mathrm{d}t}{\mathrm{d}x} \right|_{x=s} = -\frac{B}{s^2}. \tag{6.4}$$

In other words, the distribution of event sizes has precisely the power-law form seen in figure 1.4, with an exponent $\tau = 2$ which is in good agreement with the fossil data. (If we use the power-law fit to the cumulative extinction intensity suggested by Sibani et al., the exponent works out at about $\tau = 2.5$, which is outside the standard error on the value measured in the fossil record—another reason for preferring the logarithmic fit.)

There are problems with this argument. The analysis assumes that the extinction rate takes the idealized form of eq. (6.3), whereas in fact this equation represents only the average behavior of the real data. In reality, there is a great deal of fluctuation about this form. For example, eq. (6.3) implies that all the large extinction events happened in the earliest part of the fossil record, whereas in fact this is not true. The two largest events of all time (the late-Permian and end-Cretaceous events) happened in the second half of the Phanerozoic. Clearly then this analysis cannot tell the entire story.

A more serious problem is that this theory is really just "passing the buck." It doesn't tell us how, in biological terms, the observed extinction size distribution comes about. All it does is tell us that one distribution arises because of another. The extinction size distribution may be a result of the falloff in the average extinction rate, but where does the falloff come from?

The origin of the decline in the extinction rate has been a topic of debate for many years. It has been suggested that the decline may be a sampling bias in the data, arising perhaps from variation in the quality of the fossil record through geologic time (Pease 1992) or from changes in taxonomic structure (Flessa and Jablonski 1985). As with the increase in diversity discussed in section 1.2.2.3, however, many believe that these biases are not enough to account entirely for the observed extinction decline. Raup and Sepkoski (1982) have suggested instead that the decline could be the result of a slow evolutionary increase in the mean fitness of species, fitter species becoming extinct less easily than their less-fit ancestors. This appears to be a plausible suggestion, but it has a number of problems. With respect to what are we measuring fitness in this case? Do we mean fitness relative to other species? Surely not, since if all species are increasing in fitness at roughly the same rate, then their fitness relative to one another will remain approximately constant. (This is another aspect of van Valen's "Red Queen hypothesis," which we mentioned in section 1.3.) Do we then mean fitness with respect to the environment, and if so, how is such a fitness defined? The reset model attempts to address these questions and quantify the theory of increasing species fitness.

6.2   THE RESET MODEL

The basic idea of the reset model is that species are evolving on high-dimensional rugged fitness landscapes of the kind considered previously in Chapter 2. Suppose a species is evolving on such a landscape by mutations which take it from one local peak to another at approximately regular intervals of time. (This contrasts with the picture proposed by Bak and Sneppen (1993)—see section 3.1—in which the time between evolutionary jumps is not constant, but depends on a barrier variable which measures how difficult a certain jump is.) If the species moves to a new peak where the fitness is higher than the fitness at the previous peak, then the new strain will replace the old one. If the dimensionality of the landscape is sufficiently high, then the chance of a species retracing its steps and encountering the same peak twice is small and can be neglected. In this case, the process of sampling the fitness at successive peaks is equivalent to drawing a series of independent random fitness values from some fixed distribution, and keeping a record of the highest one encountered so far. Each time the current highest value is replaced by a new one, an evolutionary event has taken place in the model and such events correspond to pseudoextinction of the ancestral species. Sibani et al. refer to this process as a "resetting" of the fitness of the species (hence the name "reset model"), and to the entire dynamics of the model as a "record dynamics."

The record dynamics is simple enough to permit the calculation of distributions of a number of quantities of interest. First of all, Sibani et al. showed that the total number of evolution/extinction events happening between an initial time t_0 and a later time t goes as $\log(t - t_0)$ on average, regardless of the distribution from which the random numbers are drawn. This of course is precisely the form seen in the fossil data, eq. (6.2), and immediately implies that the number of events per unit time falls off as $1/(t - t_0)$. Then the arguments leading up to eq. (6.4) tell us that we should expect a distribution of sizes of extinction events with an exponent $\tau = 2$, as in the fossil data.

We can also calculate the distribution of the lifetimes of species. Assuming that the lifetime of a species is the interval between the evolutionary event which creates it and the next event, in which it disappears, it turns out that the reset model implies a distribution of lifetimes which is power-law in form with an exponent $\alpha = 1$, again independent of the distribution of the random numbers used. This is some way from the value $\alpha = 1.7 \pm 0.3$ observed in the fossil data (section 1.2.2.4), but no more so than for most of the other models discussed previously.

6.3 EXTINCTION MECHANISMS

The model described so far contains only a pseudoextinction mechanism; there is no true extinction taking place, a situation which we know not to be representative of the fossil record. Sibani et al. suggested an extension of their model to incorporate a true extinction mechanism based on competition between species. In this version of the model each species interacts with a number of neighboring species. Sibani et al. placed the species on a lattice and allowed each one to interact with its nearest neighbors on the lattice. (Other choices would also be possible, such as the random neighbors of the NK and Solé–Manrubia models, for instance.) If a species increases its fitness to some new value through an evolutionary event, then any neighboring species with fitness lower than this new value becomes extinct. The justification for this extinction mechanism is that neighboring species are in direct competition with one another and therefore the fitter species tends to wipe out the less-fit one by competitive exclusion. As in most of the other models we have considered, the number of species in the model is maintained at a constant level by repopulating empty niches with new species whose fitnesses are, in this case, chosen at random. Curiously, Sibani et al. did not calculate the distribution of the sizes of extinction events in this version of the model, although they did show that the new version has a steeper species lifetime distribution; it is still a power law but has an exponent of $\alpha = 2$, a value somewhat closer to the $\alpha = 1.7 \pm 0.3$ seen in the fossil data.

CHAPTER 7

Summary

In this book we have studied a large number of recent quantitative models aimed at explaining a variety of large-scale trends seen in the fossil record. These trends include the occurrence of mass extinctions, the distribution of the sizes of extinction events, the distribution of the lifetimes of taxa, the distribution of the numbers of species per genus, and the apparent decline in the average extinction rate. None of the models presented match all the fossil data perfectly, but all of them offer some suggestion of possible mechanisms which may be important to the processes of extinction and origination. In this chapter we conclude our review by briefly summarizing the properties and predictions of each of the models once more. Much of the interest in these models has focused on their ability (or lack of ability) to predict the observed values of exponents governing distributions of a number of quantities. In Table 7.1 we summarize the values of these exponents for each of the models.

Most of the models we have described attempt to provide possible explanations for a few specific observations. (1) The fossil record appears to have a power-law (i.e., scale-free) distribution of the sizes of extinction events, with an exponent close to 2 (section 1.2.2.1). (2) The distribution of the lifetimes of genera also appears to follow a power law, with exponent about 1.7 (section 1.2.2.4). (3) The number of species per genus appears to follow a power law with exponent about 1.5 (section 1.2.3.1).

TABLE 7.1   Exponents of various distributions as measured in the fossil record, and in some of the models described in this book.

	Exponent of Distribution		
	Extinction Size	Lifetime	Species/Genus
	τ	α	β
Fossil Data	**2.0 ± 0.2**	**1.7 ± 0.3**	**1.5 ± 0.1**
NKCS	$\simeq 1$	–	–
Bak and Sneppen	1 to 3/2	1	–
Vandewalle and Ausloos	1.49 ± 0.01	–	1.89 ± 0.03
Newman and Roberts	2.02 ± 0.03	–	–
Solé and Manrubia	2.05 ± 0.06	–	2.05 ± 0.06
Amaral and Meyer	1.97 ± 0.05	–	–
Newman	2.02 ± 0.02	1.03 ± 0.05	1.6 ± 0.1
Manrubia and Paczuski	1.9 ± 0.1	$\simeq 2$	$\simeq 2$
Sibani et al.	2	1	–

One of the first models to attempt an explanation of these observations was the NK model of Kauffman and co-workers. In this model, extinction is driven by coevolutionary avalanches. When tuned to the critical point between chaotic and frozen regimes, the model displays a power-law distribution of avalanche sizes with an exponent of about 1. It has been suggested that this could in turn lead to a power-law distribution of the sizes of extinction events, although the value of 1 for the exponent is not in agreement with the value 2 measured in the fossil extinction record. It is not clear by what mechanism the extinction would be produced in this model.

Building on Kauffman's ideas, Bak and Sneppen proposed a simpler model which not only produces coevolutionary avalanches, but also self-organizes to its own critical point, thereby automatically producing a power-law distribution of avalanche sizes, regardless of other parameters in the system. Again the exponent of the distribution is in the vicinity of one, which is not in agreement with the fossil record. Many extensions of the Bak–Sneppen model have been proposed. We have described the multitrait model of Boettcher and Paczuski which is less realistic but has the advantage of being exactly solvable, the model of Vandewalle and Ausloos which incorporates speciation effects and phylogenetic trees, the model of Head and Rodgers which also proposes a speciation mechanism, and the model of Newman and Roberts which introduces true extinction via environmental stress.

A different, but still biotic, extinction mechanism has been investigated by Solé and Manrubia, who proposed a "connection" model based on ideas of ecological competition. It is not clear whether ecological effects have made an important contribution to the extinction that we see in the fossil record, although the current consensus appears to be that they have not. The Solé–Manrubia model, like Kauffman's NK model, is a true critical model, which only produces power-law distributions when tuned to its critical point. Unlike Kauffman's model however, the model of Solé and Manrubia produces the correct value for the extinction size distribution when tuned to this point. We have also described two other models of extinction through ecological interaction: the food chain models of Amaral and Meyer and of Abramson.

A third distinct extinction mechanism is extinction through environmental stress, which has been investigated in modeling work by Newman. In Newman's model, species with low tolerance for stress become extinct during periods of high stress, and no species interactions are included at all. The model gives a value of 2 for the extinction size distribution, the same as that seen in the fossil record. Wilke and Martinetz have proposed a more realistic version of the same model in which recovery after mass extinctions takes place gradually, rather than instantaneously. Another related model is that of Manrubia and Paczuski in which extinction is also caused by coherent "shocks" to the ecosystem, although the biological justification for these shocks is different from that given by Newman. Their model also generates a power-law distribution of extinction sizes with exponent 2.

Finally, we have looked at the "reset model" of Sibani et al., which proposes that the distribution of sizes of extinction events is a result of declining extinction intensity during the Phanerozoic. The decline is in turn explained as a result of increasing average fitness of species as they evolve.

Clearly there are a large number of competing models here, and simply studying quantities such as the distribution of the sizes of extinction events is not going to allow us to distinguish between them. In particular, the question of whether the dominant mechanisms of extinction are biotic or abiotic is interesting and thus far undecided—the fossil data we have at present do not clearly distinguish between biotic and abiotic, although paleontological wisdom primarily leans in favor of the abiotic. However, the models we have give us a good feeling for what mechanisms might be important for generating the observed distributions. A sensible next step would be to look for signatures, in the fossil record or elsewhere, which might allow us to distinguish between these different mechanisms.

References

ABRAMSON, G. "Ecological Model of Extinctions." *Phys. Rev. E* **55** (1997): 785–788.

ADAMI, C. "Self-Organized Criticality in Living Systems." *Phys. Letts. A* **203** (1995): 29–32.

ALVAREZ, L. W. "Experimental Evidence that an Asteroid Impact Led to the Extinction of Many Species 65 Million Years Ago." *Proc. Natl. Acad. Sci.* **80** (1983): 627–642.

ALVAREZ, L. W. "Mass Extinctions Caused by Large Bolide Impacts." *Physics Today* **40** (1987): 24–33.

ALVAREZ, L. W., W. ALVAREZ, F. ASARA, AND H. V. MICHEL. "Extraterrestrial Cause for the Cretaceous–Tertiary Extinction." *Science* **208** (1980): 1095–1108.

AMARAL, L. A. N., AND M. MEYER. "Environmental Changes, Coextinction, and Patterns in the Fossil Record." *Phys. Rev. Lett.* **82** (1999): 652–655.

BAK, P. *How Nature Works: The Science of Self-Organized Criticality.* New York: Copernicus, 1996.

BAK. P., H. FLYVBJERG, AND B. LAUTRUP. "Coevolution in a Rugged Fitness Landscape." *Phys. Rev. A* **46** (1992): 6724–6730.

BAK, P., AND K. SNEPPEN. "Punctuated Equilibrium and Criticality in a Simple Model of Evolution." *Phys. Rev. Lett.* **71** (1993): 4083–4086.

BAK, P., C. TANG, AND K. WIESENFELD. "Self-Organized Criticality: An Explanation of $1/f$ Noise." *Phys. Rev. Lett.* **59** (1987): 381–384.

BENTON, M. J. "Progress and Competition in Macroevolution." *Biol. Rev.* **62** (1987): 305–338.

BENTON, M. J. "Extinction, Biotic Replacements and Clade Interactions." In *The Unity of Evolutionary Biology,* edited by E. C. Dudley. Portland: Dioscorides, 1991.

BENTON, M. J. *The Fossil Record 2.* London: Chapman and Hall, 1993.

BENTON, M. J. "Diversification and Extinction in the History of Life." *Science* **268** (1995): 52–58.

BINNEY, J. J., N. J. DOWRICK, A. J. FISHER, AND M. E. J. NEWMAN. *The Theory of Critical Phenomena.* Oxford: Oxford University Press, 1992.

BOETTCHER, S., AND M. PACZUSKI. "Exact Results for Spatiotemporal Correlation in a Self-Organized Critical Model of Punctuated Equilibrium." *Phys. Rev. Lett.* **76** (1996): 348–351.

BOURGEOIS, T., W. A. CLEMENS, R. A. SPICER, T. A. AGER, L. D. CARTER, AND W. V. SLITER. "A Tsunami Deposit at the Cretaceous–Tertiary Boundary in Texas." *Science* **241** (1988): 567–571.

BOWRING, S. A., J. P. GROTZINGER, C. E. ISACHSEN, A. H. KNOLL, S. M. PELECHATY, AND P. KOLOSOV. "Calibrating Rates of Early Cambrian Evolution." *Science* **261** (1993): 1293–1298.

BURLANDO, B. "The Fractal Dimension of Taxonomic Systems." *J. Theor. Biol.* **146** (1990): 99–114.

BURLANDO, B. "The Fractal Geometry of Evolution." *J. Theor. Biol.* **163** (1993): 161–172.

CHIAPPE, L. M. "The First 85 Million Years of Avian Evolution." *Nature* **378** (1995): 349–355.

CLEMENS, W. A. "Evolution of the Vertebrate Fauna During the Cretaceous–Tertiary Transition." In *Dynamics of Extinction,* edited by D. K. Elliott. New York: Wiley, 1986.

COURTILLOT, V., G. FERAUD, H. MALUSHI, D. VANDAMME, M. G. MOREAU, AND J. BESSE. "Deccan Flood Basalts and the Cretaceous/Tertiary Boundary." *Nature* **333** (1988): 843–846.

COURTILLOT, V., AND Y. GAUDEMER. "Effects of Mass Extinctions on Biodiversity." *Nature* **381** (1996): 146–148.

DARWIN, C. *On the Origin of Species by Means of Natural Selection.* London: John Murray, 1859. Reprint: Harvard University Press, Cambridge, MA, 1964.

DAVIS, M., P. HUT, AND R. A. MULLER. "Extinction of Species by Periodic Comet Showers." *Nature* **308** (1984): 715–717.

DE BOER, J., A. D. JACKSON, AND T. WETTIG. "Criticality in Simple Models of Evolution." *Phys. Rev. E* **51** (1995): 1059–1073.

DERRIDA, B. "Random Energy Model: The Limit of a Family of Disordered Models." *Phys. Rev. Lett.* **45** (1980): 79–82.

DERRIDA, B. "Random Energy Model: An Exactly Solvable Model of Disordered Systems." *Phys. Rev. B* **24** (1981): 2613–2626.

DROSSEL, B. "Extinction Events and Species Lifetimes in a Simple Ecological Model." *Phys. Rev. Lett.* **81** (1999): 5011–5014.

DUNCAN, R. A., AND D. G. PYLE. "Rapid Eruption of the Deccan Basalts at the Cretaceous/Tertiary Boundary." *Nature* **333** (1988): 841–843.

EBLE, G. J. "The Role of Development in Evolutionary Radiations." In *Biodiversity Dynamics: Turnover of Populations, Taxa and Communities,* edited by M. L. McKinney. New York: Columbia University Press, 1998.

EBLE, G. J. "Originations: Land and Sea Compared." *Geobios* **32** (1999): 223–234.

ELLIS, J., AND D. M. SCHRAMM. "Could a Nearby Supernova Explosion Have Caused a Mass Extinction?" *Proc. Natl. Acad. Sci.* **92** (1995): 235–238.

ERWIN, D. H. "Understanding Biotic Recoveries." In *Evolutionary Paleobiology,* edited by D. Jablonski, D. Erwin, and I. Lipps. Chicago: University of Chicago Press, 1996.

FISCHER, K. H., AND J. A. HERTZ. *Spin Glasses.* Cambridge: Cambridge University Press, 1991.

FLESSA, K. W., AND D. JABLONSKI. "Extinction is Here to Stay." *Paleobiology* **9** (1983): 315–321.

FLESSA, K. W., AND D. JABLONSKI. "Declining Phanerozoic Background Extinction Rates: Effect of Taxonomic Structure?" *Nature* **313** (1985): 216–218.

FLYVBJERG, H., K. SNEPPEN, AND P. BAK. "Mean Field Theory for a Simple Model of Evolution." *Phys. Rev. Lett.* **71** (1993): 4087–4090.

FOX, W. T. "Harmonic Analysis of Periodic Extinctions." *Paleobiology* **13** (1987): 257–271.

GAUTHIER, J. A. "Saurischian Monophyly and the Origin of Birds." *Mem. Calif. Acad. Sci.* **8** (1986): 1–47.

GILINSKY, N. L., AND R. K. BAMBAC. "Asymmetrical Patterns of Origination and Extinction in Higher Taxa." *Paleobiology* **13** (1987): 427–445.

GLEN, W. *The Mass Extinction Debates.* Stanford: Stanford University Press, 1994.

GRIEVE, R. A. F., AND E. M. SHOEMAKER. "The Record of Past Impacts on Earth." In *Hazards Due to Comets and Asteroids,* edited by T. Gehrels. Tucson: University of Arizona Press, 1994.

GRIMMETT, G. R., AND D. R. STIRZAKER. *Probability and Random Processes,* 2d ed. Oxford: Oxford University Press, 1992.

HALLAM, A. "The Case for Sea-Level Change as a Dominant Causal Factor in Mass Extinction of Marine Invertebrates." *Phil. Trans. Roy. Soc. B* **325** (1989): 437–455.

HALLOCK, P. "Why are Large Foraminifera Large?" *Paleobiology* **11** (1986): 195–208.

HARLAND, W. B., R. ARMSTRONG, V. A. COX, L. E. CRAIG, A. G. SMITH, AND D. G. SMITH. *A Geologic Time Scale 1989.* Cambridge: Cambridge University Press, 1990.

HEAD, D. A., AND G. J. RODGERS. "Speciation and Extinction in a Simple Model of Evolution." *Phys. Rev. E* **55** (1997): 3312–3319.

HERTZ, J. A., A. S. KROGH, AND R. G. PALMER. *Introduction to the Theory of Neural Computation.* Santa Fe Institute Studies in the Sciences of Complexity, Lect. Notes Vol. I. Reading: Addison-Wesley, 1991.

HOFFMAN, A. A., AND P. A. PARSONS. *Evolutionary Genetics and Environmental Stress.* Oxford: Oxford University Press, 1991.

HUT, P., W. ALVAREZ, W. P. ELDER, T. HANSEN, E. G. KAUFFMAN, G. KILLER, E. M. SHOEMAKER, AND P. R. WEISSMAN. "Comet Showers as a Cause of Mass Extinctions." *Nature* **329** (1987): 118–125.

JABLONSKI, D. "Marine Regressions and Mass Extinctions: A Test Using the Modern Biota." In *Phanerozoic Diversity Patterns,* edited by J. W. Valentine. Princeton: Princeton University Press, 1985.

JABLONSKI, D. "Background and Mass Extinctions: The Alternation of Macroevolutionary Regimes." *Science* **231** (1986): 129–133.

JABLONSKI, D. "The Biology of Mass Extinction: A Palaeontological View." *Phil. Trans. Roy. Soc. B* **325** (1989): 357–368.

JABLONSKI, D. "Extinctions: A Paleontological Perspective." *Science* **253** (1991): 754–757.

JABLONSKI, D. "The Tropics as a Source of Evolutionary Novelty through Geological Time." *Nature* **364** (1993): 142–144.

JABLONSKI, D. "Extinctions in the Fossil Record." In *Extinction Rates,* edited by J. H. Lawton and R. M. May. Oxford: Oxford University Press, 1995.

JABLONSKI, D., AND D. J. BOTTJER. "The Ecology of Evolutionary Innovation: The Fossil Record." In *Evolutionary Innovations,* edited by M. Nitecki. Chicago: University of Chicago Press, 1990.

JABLONSKI, D., AND D. J. BOTTJER. "The Origin and Diversification of Major Groups: Environmental Patterns and Macroevolutionary Lags." In *Major Evolutionary Radiations,* edited by P. D. Taylor and G. P. Larwood. Oxford: Oxford University Press, 1990.

JABLONSKI, D., AND D. J. BOTTJER. "Onshore-Offshore Trends in Marine Invertebrate Evolution." In *Causes of Evolution: A Paleontological Perspective,* edited by R. M. Ross and W. D. Allmon. Chicago: University of Chicago Press, 1990.

KAUFFMAN, S. A. *Origins of Order: Self-Organization and Selection in Evolution.* Oxford: Oxford University Press, 1993.

KAUFFMAN, S. A. *At Home in the Universe.* Oxford: Oxford University Press, 1995.

KAUFFMAN, S. A., AND J. JOHNSEN. "Coevolution to the Edge of Chaos: Coupled Fitness Landscapes, Poised States, and Coevolutionary Avalanches." *J. Theor. Biol.* **149** (1991): 467–505.

KAUFFMAN, S. A., AND S. LEVIN. "Towards a General Theory of Adaptive Walks on Rugged Landscapes." *J. Theor. Biol.* **128** (1987): 11–45.

KAUFFMAN, S. A., AND A. S. PERELSON. *Molecular Evolution on Rugged Landscapes: Proteins, RNA, and the Immune Response.* Reading: Addison–Wesley, 1990.

KAUFFMAN, S. A., AND E. W. WEINBERGER. "The NK Model of Rugged Fitness Landscapes and Its Application to Maturation of the Immune Response." *J. Theor. Biol.* **141** (1989): 211–245.

KIRCHNER, J. W., AND A. WEIL. "No Fractals in Fossil Extinction Statistics." *Nature* **395** (1998): 337–338.

KRAMER, M., N. VANDEWALLE, AND M. AUSLOOS. "Speciations and Extinction in a Self-Organizing Critical Model of Tree-like Evolution." *J. Phys. I France* **6** (1996): 599–606.

LANGTON, C. G. *Artificial Life: An Overview*. Cambridge: MIT Press, 1995.

LOPER, D. E., K. McCARTNEY, AND G. BUZYNA. "A Model of Correlated Periodicity in Magnetic-Field Reversals, Climate and Mass Extinctions." *J. Geol.* **96** (1988): 1–15.

LYELL, C. *Principles of Geology, Vol. 2*. London: Murray, 1832.

MACKEN, C. A., AND A. S. PERELSON. "Protein Evolution on Rugged Landscapes." *Proc. Natl. Acad. Sci.* **86** (1989): 6191–6195.

MANRUBIA, S. C., AND M. PACZUSKI. "A Simple Model of Large Scale Organization in Evolution." *Intl. J. Mod. Phys. C* **9** (1998): 1025–1032.

MASLOV, S., M. PACZUSKI, AND P. BAK. "Avalanches and $1/f$ Noise in Evolution and Growth Models." *Phys. Rev. Lett.* **73** (1994): 2162–2165.

MAY, R. M. "How Many Species?" *Phil. Trans. Roy. Soc. B* **330** (1990): 293–304.

MAYNARD SMITH, J. "The Causes of Extinction." *Phil. Trans. Roy. Soc. B* **325** (1989): 241–252.

MAYNARD SMITH, J., AND G. R. PRICE. "The Logic of Animal Conflict." *Nature* **246** (1973): 15–18.

McLAREN, D. J. "Detection and Significance of Mass Killings." *Hist. Biol.* **2** (1988): 5–15.

McNAMARA, K. J. "Echinoids." In *Evolutionary Trends,* edited by K. J. McNamara. London: Belhaven Press, 1990.

MITCHELL, M. *An Introduction to Genetic Algorithms*. Cambridge: MIT Press, 1996.

MONTROLL, E. W., AND M. F. SHLESINGER. "On $1/f$ Noise and Other Distributions with Long Tails." *Proc. Natl. Acad. Sci.* **79** (1982): 3380–3383.

MORRISON, D. *The Spaceguard Survey: Report of the NASA International Near-Earth Object Detection Workshop*. Pasadena: Jet Propulsion Laboratory, 1992.

NEWMAN, M. E. J. "Self-Organized Criticality, Evolution and the Fossil Extinction Record." *Proc. Roy. Soc. Lond. B* **263** (1996): 1605–1610.

NEWMAN, M. E. J. "A Model of Mass Extinction." *J. Theor. Biol.* **189** (1997): 235–252.

NEWMAN, M. E. J., AND G. J. EBLE. "Power Spectra of Extinction in the Fossil Record." *Proc. Roy. Soc. Lond. B* **266** (1999): 1267–1270.

NEWMAN, M. E. J., AND G. J. EBLE. "Decline in Extinction Rates and Scale Invariance in the Fossil Record." *Paleobiology* (1999): in press.

NEWMAN, M. E. J., S. M. FRASER, K. SNEPPEN, AND W. A. TOZIER. Comment on "Self-Organized Criticality in Living Systems." *Phys. Lett. A* **228** (1997): 201–203.

NEWMAN, M. E. J., AND B. W. ROBERTS. "Mass Extinction: Evolution and the Effects of External Influences on Unfit Species." *Proc. Roy. Soc. Lond. B* **260** (1995): 31–37.

NEWMAN, M. E. J., AND P. SIBANI. "Extinction, Diversity and Survivorship of Taxa in the Fossil Record." *Proc. Roy. Soc. Lond. B* **266** (1999): 1593–1600.

NEWMAN, M. E. J., AND K. SNEPPEN. "Avalanches, Scaling and Coherent Noise." *Phys. Rev. E* **54** (1996): 6226–6231.

PACZUSKI, M., S. MASLOV, AND P. BAK. "Avalanche Dynamics in Evolution, Growth, and Depinning Models." *Phys. Rev. E* **53** (1996): 414–443.

PANG, N. N. "The Bak–Sneppen Model: A Self-Organized Critical Model of Biological Evolution." *Intl. J. Mod. Phys. B* **11** (1997): 1411–1444.

PARSONS, P. A. "Stress, Extinctions and Evolutionary Change: From Living Organisms to Fossils." *Biol. Rev.* **68** (1993): 313–333.

PATTERSON, C., AND A. B. SMITH. "Is the Periodicity of Extinctions a Taxonomic Artifact?" *Nature* **330** (1987): 248–251.

PATTERSON, C., AND A. B. SMITH. "Periodicity in Extinction: The Role of the Systematics." *Ecology* **70** (1989): 802–811.

PATTERSON, R. T., AND A. D. FOWLER. "Evidence of Self Organization in Planktic Foraminiferal Evolution: Implications for Interconnectedness of Paleoecosystems." *Geology* **24** (1996): 215–218.

PEASE, C. M. "On the Declining Extinction and Origination Rates of Fossil Taxa." *Paleobiology* **18** (1992): 89–92.

PLOTNICK, R. E., AND M. L. MCKINNEY. "Ecosystem Organization and Extinction Dynamics." *Palaios* **8** (1993): 202–212.

RAMPINO, M. R., AND R. B. STOTHERS. "Terrestrial Mass Extinctions, Cometary Impacts and the Sun's Motion Perpendicular to the Galactic Plane." *Nature* **308** (1984): 709–712.

RAUP, D. M. "Biases in the Fossil Record of Species and Genera." *Bulletin of the Carnegie Museum of Natural History* **13** (1979): 85–91.

RAUP, D. M. "Size of the Permo-Triassic Bottleneck and Its Evolutionary Implications." *Science* **206** (1979): 217–218.

RAUP, D. M. "Magnetic Reversals and Mass Extinctions." *Nature* **314** (1985): 341–343.

RAUP, D. M. "Biological Extinction in Earth History." *Science* **231** (1986): 1528–1533.

RAUP, D. M. *Extinction: Bad Genes or Bad Luck?* New York: Norton, 1991.

RAUP, D. M. "A Kill Curve for Phanerozoic Marine Species." *Paleobiology* **17** (1991): 37–48.

RAUP, D. M. "Large-Body Impact and Extinction in the Phanerozoic." *Paleobiology* **18** (1992): 80–88.

RAUP, D. M. "Extinction Models." In *Evolutionary Paleobiology,* edited by D. Jablonski, D. H. Erwin, and J. H. Lipps. Chicago: University of Chicago Press, 1996.

RAUP, D. M., AND G. E. BOYAJIAN. "Patterns of Generic Extinction in the Fossil Record." *Paleobiology* **14** (1988): 109–125.

RAUP, D. M., AND J. J. SEPKOSKI, JR. "Mass Extinctions in the Marine Fossil Record" *Science* **215** (1982): 1501–1503.

RAUP, D. M., AND J. J. SEPKOSKI, JR. "Periodicity of Extinctions in the Geologic Past." *Proc. Natl. Acad. Sci.* **81** (1984): 801–805.

RAUP, D. M., AND J. J. SEPKOSKI, JR. "Periodic Extinctions of Families and Genera." *Science* **231** (1986): 833–836.

RAUP, D. M., AND J. J. SEPKOSKI, JR. "Testing for Periodicity of Extinction." *Science* **241** (1988): 94–96.

RAY, T. S. "An Evolutionary Approach to Synthetic Biology." *Artificial Life* **1** (1994): 179–209.

RAY, T. S. "Evolution, Complexity, Entropy and Artificial Reality." *Physica D* **75** (1994): 239–263.

ROBERTS, B. W., AND M. E. J. NEWMAN. "A Model for Evolution and Extinction." *J. Theor. Biol.* **180** (1996): 39–54.

ROSENZWEIG, M. L. *Species Diversity in Space and Time.* Cambridge: Cambridge University Press, 1995.

ROY, K. "The Roles of Mass Extinction and Biotic Interaction in Large-Scale Replacements." *Paleobiology* **22** (1996): 436–452.

SCHMOLTZI, K., AND H. G. SCHUSTER. "Introducing a Real Time Scale into the Bak–Sneppen Model." *Phys. Rev. E* **52** (1995): 5273–5280.

SEPKOSKI, J. J., JR. "Perodicity of Extinction and the Problem of Catastophism in the History of Life." *J. Geo. Soc. Lond.* **146** (1988): 7–19.

SEPKOSKI, J. J., JR. "The Taxonomic Structure of Periodic Extinction." In *Global Catastrophes in Earth History,* edited by V. L. Sharpton and P. D. Ward. *Geological Society of America Special Paper* **247** (1990): 33–44.

SEPKOSKI, J. J., JR. "Diversity in the Phanerozoic Oceans: A Partisan Review." In *The Unity of Evolutionary Biology,* edited by E. C. Dudley. Portland: Dioscorides, 1991.

SEPKOSKI, J. J., JR. "A Compendium of Fossil Marine Animal Families," 2d ed. *Milwaukee Public Museum Contributions in Biology and Geology* **83** (1993).

SEPKOSKI, J. J., JR. "Patterns of Phanerozoic Extinction: A Perspective from Global Databases." In *Global Events and Event Stratigraphy,* edited by O. H. Walliser. Berlin: Springer-Verlag, 1996.

SEPKOSKI, J. J., JR. "Rates of Speciation in the Fossil Record." *Phil. Trans. Roy. Soc. B* **353** (1998): 315–326.

SEPKOSKI, J. J., JR., AND D. C. KENDRICK. "Numerical Experiments with Model Monophyletic and Paraphyletic Taxa." *Paleobiology* **19** (1993): 168–184.

SIBANI, P., AND P. LITTLEWOOD. "Slow Dynamics from Noise Adaptation." *Phys. Rev. Lett.* **71** (1993): 1482–1485.

SIBANI, P., M. R. SCHMIDT, AND P. ALSTRØM. "Fitness Optimization and Decay of Extinction Rate through Biological Evolution." *Phys. Rev. Lett.* **75** (1995): 2055–2058.

SIBANI, P., M. R. SCHMIDT, AND P. ALSTRØM. "Evolution and Extinction Dynamics in Rugged Fitness Landscapes." *Intl. J. Mod. Phys. B* **12** (1998): 361–391.

SIGNOR, P. W., AND J. H. LIPPS. "Sampling Bias, Gradual Extinction Patterns, and Catastrophes in the Fossil Record." In *Geological Implications of Impacts of Large Asteroids and Comets on the Earth,* edited by L. T. Silver and P. H. Schultz. *Geological Society of America Special Paper* **190** (1982): 291–296.

SIMPSON, G. G. "How Many Species?" *Evolution* **6** (1952): 342–342.

SNEPPEN, K. "Extremal Dynamics and Punctuated Co-evolution." *Physica A* **221** (1995): 168–179.

SNEPPEN, K., P. BAK, H. FLYVBJERG, H., AND M. H. JENSEN. "Evolution as a Self-Organized Critical Phenomenon." *Proc. Natl. Acad. Sci.* **92** (1995): 5209–5213.

SNEPPEN, K., AND M. E. J. NEWMAN. "Coherent Noise, Scale Invariance and Intermittency in Large Systems." *Physica D* **110** (1997): 209–222.

SOLÉ, R. V. "On Macroevolution, Extinctions and Critical Phenomena." *Complexity* **1** (1996): 40–44.

SOLÉ, R. V., AND J. BASCOMPTE. "Are Critical Phenomena Relevant to Large-Scale Evolution?" *Proc. Roy. Soc. Lond. B* **263** (1996): 161–168.

SOLÉ, R. V., J. BASCOMPTE, AND S. C. MANRUBIA. "Extinction: Bad Genes or Weak Chaos?" *Proc. Roy. Soc. Lond. B* **263** (1996): 1407–1413.

SOLÉ, R. V., AND S. C. MANRUBIA. "Extinction and Self-Organized Criticality in a Model of Large-Scale Evolution." *Phys. Rev. E* **54** (1996): R42–R45.

SOLÉ, R. V., S. C. MANRUBIA, M. BENTON, AND P. BAK. "Self-Similarity of Extinction Statistics in the Fossil Record." *Nature* **388** (1997): 764–767.

SORNETTE, D., AND R. CONT. "Convergent Multiplicative Processes Repelled from Zero: Power Laws and Truncated Power Laws." *J. Phys. I France* **7** (1997): 431–444.

STANLEY, S. M. "Marine Mass Extinction: A Dominant Role for Temperature." In *Extinctions,* edited by M. H. Nitecki. Chicago: University of Chicago Press, 1984.

STANLEY, S. M. "Paleozoic Mass Extinctions: Shared Patterns Suggest Global Cooling as a Common Cause." *Am. J. Sci.* **288** (1988): 334–352.

STANLEY, S. M. "Delayed Recovery and the Spacing of Major Extinctions." *Paleobiology* **16** (1990): 401–414.

STENSETH, N. C. "Darwinian Evolution in Ecosystems: The Red Queen View." In *Evolution.* Cambridge: Cambridge University Press, 1985.

STROGATZ, S. H., AND I. STEWART. "Coupled Oscillators and Biological Synchronization." *Sci. Am.* **269** (1993): 102–109.

VAN VALEN, L. "A New Evolutionary Law." *Evol. Theory* **1** (1973): 1–30.

VANDEWALLE, N., AND M. AUSLOOS. "The Robustness of Self-Organized Criticality Against Extinctions in a Tree-like Model of Evolution." *Europhys. Lett.* **32** (1995): 613–618.

VANDEWALLE, N., AND M. AUSLOOS. "Different Universality Classes for Self-Organized Critical Models Driven by Extremal Dynamics." *Europhys. Lett.* **37** (1997): 1–6.

VERMEIJ, G. J. *Evolution as Escalation.* Princeton: Princeton University Press, 1987.

WEISBUCH, G. *Complex Systems Dynamics.* Santa Fe Institute Studies in the Sciences of Complexity, Lect. Notes Vol. II. Reading: Addison-Wesley, 1991.

WHITMIRE, D. P., AND A. A. JACKSON. "Are Periodic Mass Extinctions Driven by a Distant Solar Companion?" *Nature* **308** (1984): 713–715.

WILDE, P., AND W. B. N. BERRY. "Destabilization of the Oceanic Density Structure and Its Significance to Marine Extinction Events." *Palaeogeog. Palaeoclimatol. Palaeoecol.* **48** (1984): 142–162.

WILKE, C., S. ALTMEYER, AND T. MARTINETZ. "Aftershocks in Coherent-Noise Models." *Physica D* **120** (1998): 401–417.

WILKE, C., AND T. MARTINETZ. "Simple Model of Evolution with Variable System Size." *Phys. Rev. E* **56** (1997): 7128–7131.

WILLIAMS, C. B. "Some Applications of the Logarithmic Series and the Index of Diversity to Ecological Problems." *J. Ecol.* **32** (1944): 1–44.

WILLIS, J. C. *Age and Area.* Cambridge: Cambridge University Press, 1922.

ZIPF, G. K. *Human Behavior and the Principle of Least Effort.* Reading: Addison-Wesley, 1949.

Index